Haynes
THE BOOK
®

AUTOMOTIVE

DIESEL
Manual

Martynn Randall

ISBN **1 84425 174 8**

British Library Cataloguing in Publication Data
A catalogue record for this book is available from the British Library.

Printed in the USA

Haynes Publishing
Sparkford, Yeovil, Somerset BA22 7JJ, England

Haynes North America, Inc
861 Lawrence Drive, Newbury Park, California 91320, USA

Editions Haynes
4, Rue de l'Abreuvoir
92415 COURBEVOIE CEDEX, France

Haynes Publishing Nordiska AB
Box 1504, 751 45 UPPSALA, Sverige

Acknowledgements

Thanks are due to the various vehicle and equipment manufacturers and importers for providing technical literature, data and illustrations for this book. These include Lucas CAV Limited and Robert Bosch Limited for the use of their illustrations.

This book is not a direct reproduction of the vehicle manufacturers' data, and its publication should not be taken as implying any technical approval by the vehicle manufacturers or importers.

We take great pride in the accuracy of information given in this manual, but vehicle manufacturers make alterations and design changes during the production run of a particular vehicle of which they do not inform us. No liability can be accepted by the authors or publishers for loss, damage or injury caused by any errors in, or omissions from, the information given.

This manual provides descriptions and explanations of most modern diesel engine systems and their components. The book is intended for the DIY mechanic.

The manual is divided into six main Chapters:

Chapter 1 provides a basic introduction to the diesel engine and its associated systems.
Chapter 2 describes maintenance and servicing operations which are unique to, or particularly important on, diesel engines.
Chapter 3 describes the components of diesel fuel systems, and provides guidelines for their replacement.
Chapter 4 provides servicing data for most diesel engines from 1990 onwards.
Chapter 5 provides fault diagnosis charts and notes.
Chapter 6 looks at the tools and equipment needed for maintenance, diagnosis and repair.

General

Although this book does not provide maintenance schedules and detailed maintenance procedures, the following points should be taken into account when working through the vehicle manufacturer's maintenance schedule.

Maintenance intervals

When both time and mileage intervals are specified by the manufacturer, the time interval should be followed if the specified mileage is not covered within the time stated. This is necessary because some fluids and systems deteriorate with time as well as with use. In particular, water trap draining and fuel filter changes should not be neglected on low-mileage vehicles. More water may accumulate in the fuel system of a vehicle which stands idle for long periods than in one which is in constant use. If water or dirt get past the fuel filter and into the injection system, serious damage may result. A clean filter is also less likely to suffer from waxing in cold weather.

Oil change intervals tend to be shorter for diesel engines than for comparable petrol engines, because more contamination and fuel dilution of the oil occurs in the diesel. Sulphur compounds in diesel fuel are particularly detrimental to the oil; if fuel with a higher than normal sulphur content has to be used, oil change intervals must be reduced.

Adverse operating conditions

Vehicles used under adverse conditions may required more frequent maintenance. 'Adverse conditions' include the following:

Mainly short journeys
Full-time towing or taxi work
Operating in extremely hot or cold climates
Driving on unmade roads or in dusty conditions
Use of inferior quality fuel

Timing belt renewal

When a toothed belt is used to drive the camshaft and/or injection pump, periodic renewal is normally specified. *If a camshaft drivebelt breaks or slips in service, extensive engine damage will almost certainly result from the ensuing piston/valve contact.* Observe the specified intervals for inspection and renewal, even if the belt appears to be in good condition. Renew a belt which is obviously frayed, or which has been contaminated with oil or fuel, without question. Renew idler or tensioner rollers at the same time if they show shake or roughness when spun, and the sprockets if they are damaged.

Cooling system maintenance

Unless otherwise specified, the coolant antifreeze concentration should be checked at the beginning of each winter, and made good if necessary. Coolant should generally be renewed every two years, in order to maintain its corrosion-inhibiting qualities; note however, that some manufacturers claim to have a 'sealed-for-life' cooling system, often filled with their own brand of coolant (this coolant may not be compatible with other brands).

After draining the old coolant, take the opportunity to flush the system if necessary, and renew any hoses which are not in good condition.

Recommended lubricants and fluids

The following are general recommendations only. Observe the vehicle manufacturer's specifications when they differ from those given here.

Engine oil

The properties necessary in an oil for diesel engines are not identical to those needed for petrol engines. This is due to the higher mechanical loads imposed by compression ignition, and to the different effects of unburnt fuel and combustion products on the oil. When a turbocharger is fitted, the oil must also be able to cope with extremely high temperatures and rotational speeds.

For temperate climates, most manufacturers specify the use of multigrade engine oil to API CE, CCMC PD2/D4, ACEA B3-96, or equivalent (or higher) ratings.

API (American Petroleum Institute) ratings show the performance of the oil for both petrol and diesel applications. Petrol ratings begin with the letter 'S' for spark ignition, and diesel with 'C' for compression ignition. The second letter denotes the rating, with 'A' being the lowest. The higher the second letter in the alphabet, the better the rating.

CCMC (Constructors' Committee of the Common Market) ratings fall into three categories; 'G' for gasoline (petrol), 'D' for commercial diesel, and 'PD' for passenger diesel. Each rating is followed by a number. The higher the number, the better the rating.

ACEA (*Association des Constructeurs European d'Automobiles*) ratings also fall into three categories 'A' for petrol engines, 'B' for 'light-duty' diesel engines, and 'E' for 'heavy-duty' diesel engines. ACEA standards replaced CCMC standards in Europe from January 1st 1996, and hence the ratings include '96', eg, 'B1-96' (this year code is likely to be updated in the future). Each rating letter is followed by a number, and the higher the number, the better the rating.

Coolant

Modern engines often expose the coolant to several different metals – for instance iron, aluminium and copper – which in the presence of plain water will interact and rust or corrode rapidly. For this reason, it is essential that the coolant contains a corrosion inhibitor, even when freezing conditions are not expected. When hard water is used in the cooling system, a scale inhibitor is also required. The corrosion and scale inhibitors lose their effectiveness after a while, so coolant must be renewed periodically – typically every two years.

Antifreeze with a methanol content is particularly to be avoided. Methanol does lower the freezing point, but is highly poisonous and inflammable; it also tends to evaporate in use, so reducing the level of protection.

Some vehicle manufacturers (notably VW/Audi) use their own brand of antifreeze in vehicles when new. Certain of these antifreeze products will not mix with other brands, so the vehicle manufacturer's recommendations should always be followed when renewing antifreeze or topping up.

Working on your car can be dangerous. This page shows just some of the potential risks and hazards, with the aim of creating a safety-conscious attitude.

General hazards

Scalding

• Don't remove the radiator or expansion tank cap while the engine is hot.
• Engine oil, automatic transmission fluid or power steering fluid may also be dangerously hot if the engine has recently been running.

Burning

• Beware of burns from the exhaust system and from any part of the engine. Brake discs and drums can also be extremely hot immediately after use.

Crushing

• When working under or near a raised vehicle, always supplement the jack with axle stands, or use drive-on ramps. *Never venture under a car which is only supported by a jack.*

• Take care if loosening or tightening high-torque nuts when the vehicle is on stands. Initial loosening and final tightening should be done with the wheels on the ground.

Fire

• Fuel is highly flammable; fuel vapour is explosive.
• Don't let fuel spill onto a hot engine.
• Do not smoke or allow naked lights (including pilot lights) anywhere near a vehicle being worked on. Also beware of creating sparks (electrically or by use of tools).
• Fuel vapour is heavier than air, so don't work on the fuel system with the vehicle over an inspection pit.
• Another cause of fire is an electrical overload or short-circuit. Take care when repairing or modifying the vehicle wiring.
• Keep a fire extinguisher handy, of a type suitable for use on fuel and electrical fires.

Electric shock

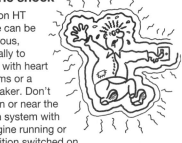

• Ignition HT voltage can be dangerous, especially to people with heart problems or a pacemaker. Don't work on or near the ignition system with the engine running or the ignition switched on.

• Mains voltage is also dangerous. Make sure that any mains-operated equipment is correctly earthed. Mains power points should be protected by a residual current device (RCD) circuit breaker.

Fume or gas intoxication

• Exhaust fumes are poisonous; they often contain carbon monoxide, which is rapidly fatal if inhaled. Never run the engine in a confined space such as a garage with the doors shut.
• Fuel vapour is also poisonous, as are the vapours from some cleaning solvents and paint thinners.

Poisonous or irritant substances

• Avoid skin contact with battery acid and with any fuel, fluid or lubricant, especially antifreeze, brake hydraulic fluid and Diesel fuel. Don't syphon them by mouth. If such a substance is swallowed or gets into the eyes, seek medical advice.
• Prolonged contact with used engine oil can cause skin cancer. Wear gloves or use a barrier cream if necessary. Change out of oil-soaked clothes and do not keep oily rags in your pocket.
• Air conditioning refrigerant forms a poisonous gas if exposed to a naked flame (including a cigarette). It can also cause skin burns on contact.

Asbestos

• Asbestos dust can cause cancer if inhaled or swallowed. Asbestos may be found in gaskets and in brake and clutch linings. When dealing with such components it is safest to assume that they contain asbestos.

Special hazards

Hydrofluoric acid

• This extremely corrosive acid is formed when certain types of synthetic rubber, found in some O-rings, oil seals, fuel hoses etc, are exposed to temperatures above 400°C. The rubber changes into a charred or sticky substance containing the acid. *Once formed, the acid remains dangerous for years. If it gets onto the skin, it may be necessary to amputate the limb concerned.*
• When dealing with a vehicle which has suffered a fire, or with components salvaged from such a vehicle, wear protective gloves and discard them after use.

The battery

• Batteries contain sulphuric acid, which attacks clothing, eyes and skin. Take care when topping-up or carrying the battery.
• The hydrogen gas given off by the battery is highly explosive. Never cause a spark or allow a naked light nearby. Be careful when connecting and disconnecting battery chargers or jump leads.

Air bags

• Air bags can cause injury if they go off accidentally. Take care when removing the steering wheel and/or facia. Special storage instructions may apply.

Diesel injection equipment

• Diesel injection pumps supply fuel at very high pressure. Take care when working on the fuel injectors and fuel pipes.

⚠ *Warning: Never expose the hands, face or any other part of the body to injector spray; the fuel can penetrate the skin with potentially fatal results.*

Remember...

DO

• Do use eye protection when using power tools, and when working under the vehicle.

• Do wear gloves or use barrier cream to protect your hands when necessary.

• Do get someone to check periodically that all is well when working alone on the vehicle.

• Do keep loose clothing and long hair well out of the way of moving mechanical parts.

• Do remove rings, wristwatch etc, before working on the vehicle – especially the electrical system.

• Do ensure that any lifting or jacking equipment has a safe working load rating adequate for the job.

DON'T

• Don't attempt to lift a heavy component which may be beyond your capability – get assistance.

• Don't rush to finish a job, or take unverified short cuts.

• Don't use ill-fitting tools which may slip and cause injury.

• Don't leave tools or parts lying around where someone can trip over them. Mop up oil and fuel spills at once.

• Don't allow children or pets to play in or near a vehicle being worked on.

General repair procedures

Whenever servicing, repair or overhaul work is carried out on the car or its components, observe the following procedures and instructions. This will assist in carrying out the operation efficiently and to a professional standard of workmanship.

Joint mating faces and gaskets

When separating components at their mating faces, never insert screwdrivers or similar implements into the joint between the faces in order to prise them apart. This can cause severe damage which results in oil leaks, coolant leaks, etc upon reassembly. Separation is usually achieved by tapping along the joint with a soft-faced hammer in order to break the seal. However, note that this method may not be suitable where dowels are used for component location.

Where a gasket is used between the mating faces of two components, a new one must be fitted on reassembly; fit it dry unless otherwise stated in the repair procedure. Make sure that the mating faces are clean and dry, with all traces of old gasket removed. When cleaning a joint face, use a tool which is unlikely to score or damage the face, and remove any burrs or nicks with an oilstone or fine file.

Make sure that tapped holes are cleaned with a pipe cleaner, and keep them free of jointing compound, if this is being used, unless specifically instructed otherwise.

Ensure that all orifices, channels or pipes are clear, and blow through them, preferably using compressed air.

Oil seals

Oil seals can be removed by levering them out with a wide flat-bladed screwdriver or similar implement. Alternatively, a number of self-tapping screws may be screwed into the seal, and these used as a purchase for pliers or some similar device in order to pull the seal free.

Whenever an oil seal is removed from its working location, either individually or as part of an assembly, it should be renewed.

The very fine sealing lip of the seal is easily damaged, and will not seal if the surface it contacts is not completely clean and free from scratches, nicks or grooves. If the original sealing surface of the component cannot be restored, and the manufacturer has not made provision for slight relocation of the seal relative to the sealing surface, the component should be renewed.

Protect the lips of the seal from any surface which may damage them in the course of fitting. Use tape or a conical sleeve where possible. Lubricate the seal lips with oil before fitting and, on dual-lipped seals, fill the space between the lips with grease.

Unless otherwise stated, oil seals must be fitted with their sealing lips toward the lubricant to be sealed.

Use a tubular drift or block of wood of the appropriate size to install the seal and, if the seal housing is shouldered, drive the seal down to the shoulder. If the seal housing is unshouldered, the seal should be fitted with its face flush with the housing top face (unless otherwise instructed).

Screw threads and fastenings

Seized nuts, bolts and screws are quite a common occurrence where corrosion has set in, and the use of penetrating oil or releasing fluid will often overcome this problem if the offending item is soaked for a while before attempting to release it. The use of an impact driver may also provide a means of releasing such stubborn fastening devices, when used in conjunction with the appropriate screwdriver bit or socket. If none of these methods works, it may be necessary to resort to the careful application of heat, or the use of a hacksaw or nut splitter device.

Studs are usually removed by locking two nuts together on the threaded part, and then using a spanner on the lower nut to unscrew the stud. Studs or bolts which have broken off below the surface of the component in which they are mounted can sometimes be removed using a stud extractor. Always ensure that a blind tapped hole is completely free from oil, grease, water or other fluid before installing the bolt or stud. Failure to do this could cause the housing to crack due to the hydraulic action of the bolt or stud as it is screwed in.

When tightening a castellated nut to accept a split pin, tighten the nut to the specified torque, where applicable, and then tighten further to the next split pin hole. Never slacken the nut to align the split pin hole, unless stated in the repair procedure.

When checking or retightening a nut or bolt to a specified torque setting, slacken the nut or bolt by a quarter of a turn, and then retighten to the specified setting. However, this should not be attempted where angular tightening has been used.

For some screw fastenings, notably cylinder head bolts or nuts, torque wrench settings are no longer specified for the latter stages of tightening, "angle-tightening" being called up instead. Typically, a fairly low torque wrench setting will be applied to the bolts/nuts in the correct sequence, followed by one or more stages of tightening through specified angles.

Locknuts, locktabs and washers

Any fastening which will rotate against a component or housing during tightening should always have a washer between it and the relevant component or housing.

Spring or split washers should always be renewed when they are used to lock a critical component such as a big-end bearing retaining bolt or nut. Locktabs which are folded over to retain a nut or bolt should always be renewed.

Self-locking nuts can be re-used in non-critical areas, providing resistance can be felt when the locking portion passes over the bolt or stud thread. However, it should be noted that self-locking stiffnuts tend to lose their effectiveness after long periods of use, and should then be renewed as a matter of course.

Split pins must always be replaced with new ones of the correct size for the hole.

When thread-locking compound is found on the threads of a fastener which is to be re-used, it should be cleaned off with a wire brush and solvent, and fresh compound applied on reassembly.

Special tools

Some repair procedures in this manual entail the use of special tools such as a press, two or three-legged pullers, spring compressors, etc. Wherever possible, suitable readily-available alternatives to the manufacturer's special tools are described, and are shown in use. In some instances, where no alternative is possible, it has been necessary to resort to the use of a manufacturer's tool, and this has been done for reasons of safety as well as the efficient completion of the repair operation. Unless you are highly-skilled and have a thorough understanding of the procedures described, never attempt to bypass the use of any special tool when the procedure described specifies its use. Not only is there a very great risk of personal injury, but expensive damage could be caused to the components involved.

Environmental considerations

When disposing of used engine oil, brake fluid, antifreeze, etc, give due consideration to any detrimental environmental effects. Do not, for instance, pour any of the above liquids down drains into the general sewage system, or onto the ground to soak away. Many local council refuse tips provide a facility for waste oil disposal, as do some garages. If none of these facilities are available, consult your local Environmental Health Department, or the National Rivers Authority, for further advice.

With the universal tightening-up of legislation regarding the emission of environmentally-harmful substances from motor vehicles, most vehicles have tamperproof devices fitted to the main adjustment points of the fuel system. These devices are primarily designed to prevent unqualified persons from adjusting the fuel/air mixture, with the chance of a consequent increase in toxic emissions. If such devices are found during servicing or overhaul, they should, wherever possible, be renewed or refitted in accordance with the manufacturer's requirements or current legislation.

OIL CARE

FOLLOW THE CODE

OIL BANK LINE

0800 66 33 66

www.oilbankline.org.uk

Note: It is antisocial and illegal to dump oil down the drain. To find the location of your local oil recycling bank, call this number free.

Length (distance)

Inches (in)	x 25.4	= Millimetres (mm)	x 0.0394	=	Inches (in)
Feet (ft)	x 0.305	= Metres (m)	x 3.281	=	Feet (ft)
Miles	x 1.609	= Kilometres (km)	x 0.621	=	Miles

Volume (capacity)

Cubic inches (cu in; in³)	x 16.387	= Cubic centimetres (cc; cm³)	x 0.061	=	Cubic inches (cu in; in³)
Imperial pints (Imp pt)	x 0.568	= Litres (l)	x 1.76	=	Imperial pints (Imp pt)
Imperial quarts (Imp qt)	x 1.137	= Litres (l)	x 0.88	=	Imperial quarts (Imp qt)
Imperial quarts (Imp qt)	x 1.201	= US quarts (US qt)	x 0.833	=	Imperial quarts (Imp qt)
US quarts (US qt)	x 0.946	= Litres (l)	x 1.057	=	US quarts (US qt)
Imperial gallons (Imp gal)	x 4.546	= Litres (l)	x 0.22	=	Imperial gallons (Imp gal)
Imperial gallons (Imp gal)	x 1.201	= US gallons (US gal)	x 0.833	=	Imperial gallons (Imp gal)
US gallons (US gal)	x 3.785	= Litres (l)	x 0.264	=	US gallons (US gal)

Mass (weight)

Ounces (oz)	x 28.35	= Grams (g)	x 0.035	=	Ounces (oz)
Pounds (lb)	x 0.454	= Kilograms (kg)	x 2.205	=	Pounds (lb)

Force

Ounces-force (ozf; oz)	x 0.278	= Newtons (N)	x 3.6	=	Ounces-force (ozf; oz)
Pounds-force (lbf; lb)	x 4.448	= Newtons (N)	x 0.225	=	Pounds-force (lbf; lb)
Newtons (N)	x 0.1	= Kilograms-force (kgf; kg)	x 9.81	=	Newtons (N)

Pressure

Pounds-force per square inch (psi; lbf/in²; lb/in²)	x 0.070	= Kilograms-force per square centimetre (kgf/cm²; kg/cm²)	x 14.223	=	Pounds-force per square inch (psi; lbf/in²; lb/in²)
Pounds-force per square inch (psi; lbf/in²; lb/in²)	x 0.068	= Atmospheres (atm)	x 14.696	=	Pounds-force per square inch (psi; lbf/in²; lb/in²)
Pounds-force per square inch (psi; lbf/in²; lb/in²)	x 0.069	= Bars	x 14.5	=	Pounds-force per square inch (psi; lbf/in²; lb/in²)
Pounds-force per square inch (psi; lbf/in²; lb/in²)	x 6.895	= Kilopascals (kPa)	x 0.145	=	Pounds-force per square inch (psi; lbf/in²; lb/in²)
Kilopascals (kPa)	x 0.01	= Kilograms-force per square centimetre (kgf/cm²; kg/cm²)	x 98.1	=	Kilopascals (kPa)
Millibar (mbar)	x 100	= Pascals (Pa)	x 0.01	=	Millibar (mbar)
Millibar (mbar)	x 0.0145	= Pounds-force per square inch (psi; lbf/in²; lb/in²)	x 68.947	=	Millibar (mbar)
Millibar (mbar)	x 0.75	= Millimetres of mercury (mmHg)	x 1.333	=	Millibar (mbar)
Millibar (mbar)	x 0.401	= Inches of water (inH₂O)	x 2.491	=	Millibar (mbar)
Millimetres of mercury (mmHg)	x 0.535	= Inches of water (inH₂O)	x 1.868	=	Millimetres of mercury (mmHg)
Inches of water (inH₂O)	x 0.036	= Pounds-force per square inch (psi; lbf/in²; lb/in²)	x 27.68	=	Inches of water (inH₂O)

Torque (moment of force)

Pounds-force inches (lbf in; lb in)	x 1.152	= Kilograms-force centimetre (kgf cm; kg cm)	x 0.868	=	Pounds-force inches (lbf in; lb in)
Pounds-force inches (lbf in; lb in)	x 0.113	= Newton metres (Nm)	x 8.85	=	Pounds-force inches (lbf in; lb in)
Pounds-force inches (lbf in; lb in)	x 0.083	= Pounds-force feet (lbf ft; lb ft)	x 12	=	Pounds-force inches (lbf in; lb in)
Pounds-force feet (lbf ft; lb ft)	x 0.138	= Kilograms-force metres (kgf m; kg m)	x 7.233	=	Pounds-force feet (lbf ft; lb ft)
Pounds-force feet (lbf ft; lb ft)	x 1.356	= Newton metres (Nm)	x 0.738	=	Pounds-force feet (lbf ft; lb ft)
Newton metres (Nm)	x 0.102	= Kilograms-force metres (kgf m; kg m)	x 9.804	=	Newton metres (Nm)

Power

Horsepower (hp)	x 745.7	= Watts (W)	x 0.0013	=	Horsepower (hp)

Velocity (speed)

Miles per hour (miles/hr; mph)	x 1.609	= Kilometres per hour (km/hr; kph)	x 0.621	=	Miles per hour (miles/hr; mph)

Fuel consumption*

Miles per gallon, Imperial (mpg)	x 0.354	= Kilometres per litre (km/l)	x 2.825	=	Miles per gallon, Imperial (mpg)
Miles per gallon, US (mpg)	x 0.425	= Kilometres per litre (km/l)	x 2.352	=	Miles per gallon, US (mpg)

Temperature

Degrees Fahrenheit = (°C x 1.8) + 32   Degrees Celsius (Degrees Centigrade; °C) = (°F - 32) x 0.56

* It is common practice to convert from miles per gallon (mpg) to litres/100 kilometres (l/100km), where mpg x l/100 km = 282

Diesel engines and injection systems

1

1   History

Rudolf Diesel invented the first commercially-successful compression-ignition engine at the end of the 19th century. Compared with the spark ignition engine, the diesel had the advantages of lower fuel consumption, the ability to use cheaper fuel, and the potential for much higher power outputs. Over the following two or three decades, such engines were widely adopted for stationary and marine applications, but the fuel injection systems used were not capable of high-speed operation. This speed limitation, and the considerable weight of the air compressor needed to operate the injection equipment, made the first diesel engines unsuitable for use in road-going vehicles.

In the 1920s, the German engineer Robert Bosch developed the in-line injection pump, a device which is still in extensive use today. The use of hydraulic systems to pressurise and inject the fuel did away with the need for a separate air compressor, and made possible much higher operating speeds. The so called 'high-speed' diesel engine became increasingly popular as a power source for goods and public transport vehicles, but for a number of reasons (including specific power output, flexibility and cheapness of manufacture), the spark-ignition engine continued to dominate the passenger car and light commercial market.

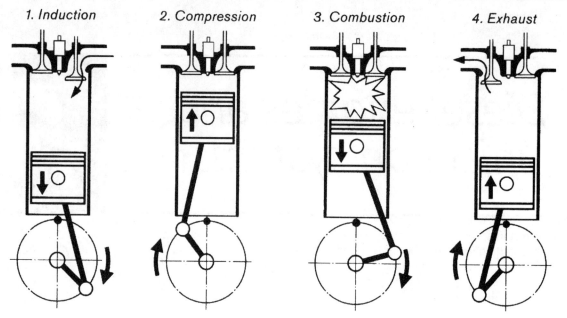

Four-stroke diesel cycle
© Robert Bosch Limited

In the 1950s and 60s, diesel engines became increasingly popular for use in taxis and vans, but it was not until the sharp rises in oil prices in the 1970s that serious attention was paid to the small passenger car market. VW's introduction of the diesel-powered Golf at the end of 1977 marked the arrival of the first 'user-friendly' diesel car, designed specifically to be acceptable to drivers who would not previously have considered abandoning the petrol engine. The diesel engine fitted to the Golf used indirect injection and a distributor type pump, and was comparable in performance to the smaller petrol engines fitted to the range.

Subsequent years have seen the growing popularity of the small diesel engine in cars and light commercial vehicles, not only for reasons of fuel economy and longevity, but also for environmental reasons. Every major European car manufacturer now offers at least one diesel-engined model. The diesel's penetration of the UK market has been relatively slow (due in part to the lack of any considerable fuel price differential in favour of diesel which exists in other parts of Europe), but it has now gained widespread acceptance, and this trend looks set to continue.

2 Principles of operation

All the diesel engines covered in this book operate on the familiar four-stroke cycle of induction, compression, combustion and exhaust *(see illustration)*. (Two-stroke diesels do exist, and may in future become important, but they are used in few light vehicles at present.) Most have four cylinders, some larger engines have six, and five- and three-cylinder engines also exist.

Induction and ignition

The main difference between diesel and petrol engines is the means by which the fuel/air mixture is introduced into the cylinder and then ignited. In the petrol engine, the fuel is mixed with the incoming air before it enters the cylinder, and the mixture is then ignited at the appropriate moment by a spark plug. At all conditions except full-throttle, the throttle butterfly restricts the airflow, and cylinder filling is incomplete.

In the diesel engine, air alone is drawn into the cylinder and then compressed. Because of the diesel's high compression ratio (typically 20:1), the air gets very hot when compressed – up to 750°C (1382°F). As the piston approaches the end of the compression stroke, fuel is injected into the combustion chamber under very high pressure, in the form of a finely-atomised spray. The temperature of the air is high enough to ignite the injected fuel as it mixes with the air. The mixture then burns and provides the energy which drives the piston downwards on the combustion (power) stroke.

When starting the engine from cold, the temperature of the compressed air in the cylinders may not be high enough to ignite the fuel. The preheating system overcomes this problem. Most modern engines have automatically-controlled preheating systems, using electric heater plugs (glow plugs) which heat the air in the combustion chamber just before and during start-up. Full details of these systems are given in Chapter 3.

On most diesel engines there is no throttle valve in the inlet tract; exceptions to this are those few engines which use a pneumatic governor, which depends on a manifold depression being created. Even more rarely, a throttle valve may be used to create manifold depression for the operation of a brake servo, though it is more usual for a separate vacuum pump to be fitted for this purpose.

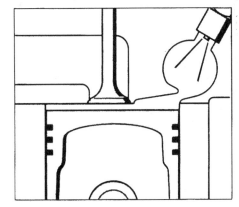

Direct and indirect injection

In practice, it is difficult to achieve smooth combustion in a small-displacement engine by injecting the fuel directly into the combustion chamber. To get around this problem, the technique of indirect injection is widely used. With indirect injection, the fuel is injected into a pre-combustion or 'swirl' chamber in the cylinder head, alongside the main combustion chamber. During the compression stroke the compressed hot air is forced into the swirl chamber where it enters a rapid swirling motion; fuel is injected into the swirl chamber, where it mixes with the rapidly moving air, enabling smoother combustion in the main combustion chamber (see illustration).

Generally speaking, indirect injection engines are less efficient than direct injection engines, and also require more preheating when starting from cold, but these disadvantages are offset by smoother and quieter operation. Until recently, direct injection engines were mostly fitted to light commercial vehicles, where increased noise and harshness are considered acceptable trade-offs for improved fuel economy. Recently, the use of electronic diesel engine control systems has allowed the development of more refined direct injection engines, and their use in passenger vehicles is now almost universal.

Indirect injection into a swirl chamber

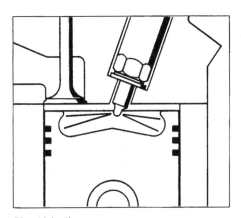

Direct injection

Indirect and direct injection
© Robert Bosch Limited

Mechanical construction

Due to the high compression ratio required in a diesel engine, and the combustion characteristics, it is necessary to ensure that the lower face of the cylinder head is flat. This is achieved by positioning the valves vertically in the cylinder head (ie, with their stems at right-angles to the cylinder head lower face), and machining the combustion chambers directly into the tops of the pistons (see illustration). Locating the combustion chambers in the pistons also enables the combustion process to be contained, and allows fine control of the combustion chamber size and shape during manufacture (all the combustion chambers in a diesel engine must be of similar size and shape).

The pistons, crankshaft and bearings of a diesel engine are generally of more robust construction than in a petrol engine of comparable size, because of the greater loads imposed by the higher compression ratio and the nature of the combustion process. This is one reason for the diesel engine's longer life. Other reasons include the lubricating qualities of diesel fuel on the cylinder bores, and the fact that the diesel engine is generally lower-revving than its petrol counterpart, having much better low-speed torque characteristics and a lower maximum speed.

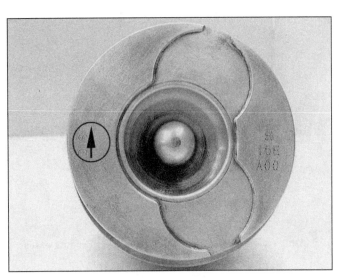

**The combustion chamber is machined into the piston crown.
Arrow points towards timing belt end**

Turbocharging

Turbochargers have long been used on large diesel engines, and are becoming common on small ones. The turbocharger uses the energy of the escaping exhaust gas to drive a turbine which pressurises the air in the inlet manifold. The air is forced into the cylinders instead of being simply sucked in. If more air is present, more fuel can be burnt and more power can be developed from the same size engine *(see illustration)*.

Greater benefit can be gained from turbocharging if the pressurised air is cooled before it enters the engine. This is done using an air-to-air heat exchanger called an intercooler. The cooled air is denser and contains more oxygen in a given volume than warm air straight from the turbocharger *(see illustrations)*.

Later engines, particularly from the VAG (Volkswagen Audi Group), may utilise a variable geometry turbocharger *(see illustration)*. With this design, the exhaust gasses entering the turbocharger pass through a variable size venture. When the engine speed is low and the gas speed is slow, the venture diameter is reduced. This has the effect of speeding up the gasses just before they meet the turbocharger wheel. This maintains a high turbine speed, which improves turbocharger performance at low speed. At higher engine speeds when the gasses are moving much faster, the venturi is enlarged. Using this system allows the turbocharger to operate at close to maximum efficiency at a greater range of engine speeds.

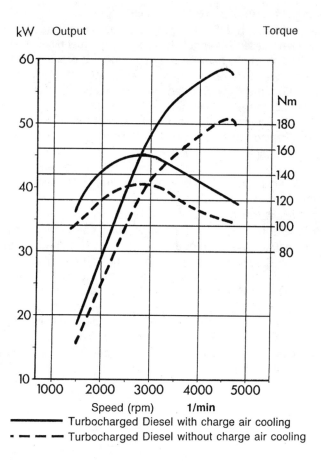

───── Turbocharged Diesel with charge air cooling
─ ─ ─ Turbocharged Diesel without charge air cooling

Engine power and torque with and without charge air cooling (intercooling)

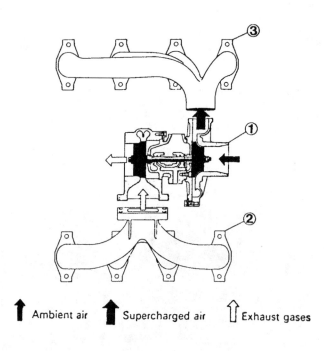

▲ Ambient air ▲ Supercharged air ⬆ Exhaust gases

Principle of turbocharging

1 Turbocharger   2 Exhaust manifold   3 Inlet manifold

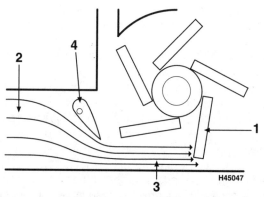

Variable geometry turbocharger at low engine speed

1 Turbine wheel   *3 High speed exhaust gasses*
2 Low speed exhaust gasses   *4 Adjustable vane*

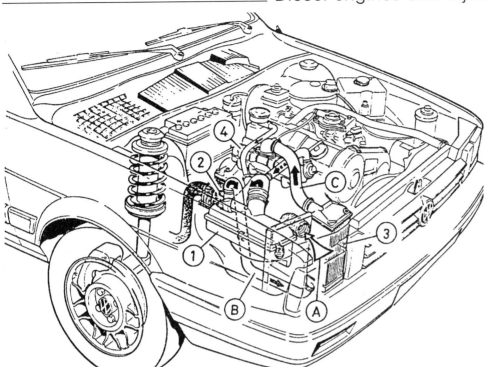

Induction airflow in a turbocharged engine with charge air cooling

1 Air cleaner
2 Turbocharger
3 Intercooler
4 Inlet manifold
A Inducted air
B Compressed air before cooling
C Compressed air after cooling

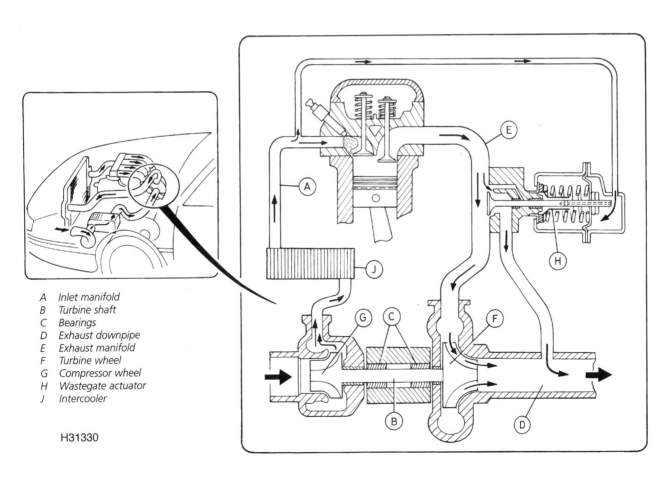

A Inlet manifold
B Turbine shaft
C Bearings
D Exhaust downpipe
E Exhaust manifold
F Turbine wheel
G Compressor wheel
H Wastegate actuator
J Intercooler

H31330

Turbocharger location and schematic view of operation – Citroën XM

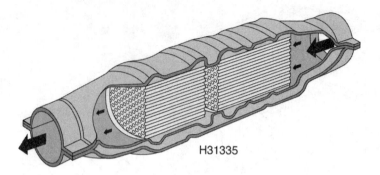

CARBON MONOXIDE (CO)
HYDROCARBONS (HC)
NITROGEN OXIDES (NO$_x$)

CARBON DIOXIDE (CO$_2$)
WATER (H$_2$O)
NITROGEN (N)

H31335

Cross-section of a typical catalytic converter

Exhaust emissions

Because combustion in a correctly-functioning diesel engine nearly always occurs in conditions of excess oxygen, there is little or no carbon monoxide (CO) present in the exhaust gas. A further environmental benefit is that there is no added lead in diesel fuel.

For many years, there was no need for complicated emission control systems on diesel engines. In the last few years however, simple catalytic converters, and exhaust gas recirculation systems, have become standard on most diesel engines in order to meet the increasingly stringent emission regulations. The advent of electronic diesel engine control systems has also helped to improve diesel engine emissions.

Catalytic converter

The catalytic converter consists of a canister containing a fine mesh impregnated with a catalyst material, over which the exhaust gases pass. The catalyst speeds up the oxidation of harmful carbon monoxide and unburnt hydrocarbons, effectively reducing the quantity of harmful products reaching the atmosphere *(see illustration)*. Because unburnt hydrocarbons contribute to particle emission, this can also be reduced to a limited extent by a catalytic converter.

A closed-loop catalytic converter system using an oxygen sensor, similar to that used on petrol engines, cannot be used on a diesel engine because a diesel engine always operates with excess air, and hence oxygen, in the exhaust gas.

Particle filter systems

Particle filters and traps reduce the level of smoke particles released into the air by the diesel engine exhaust. This is an evolving technology; not all vehicle manufacturers use particle filters, but they still manage to satisfy the 'Euro 4' emission standards (mandatory in 2005).

The particle filter used by Peugeot and Citroën (PSA) is of the soot burn-off type. It works as follows *(see illustration)*.

The soot particles are trapped in a block of filter material. The filter is carefully designed to allow the exhaust gases to flow through it, whilst trapping the soot particles; however there obviously comes a point when the filter will become blocked due to the large number of particles trapped. As the trapped particles in the filter build up, there will be a resistance to the exhaust gas flow. Pressure sensors on either side of the filter detect this resistance; when it exceeds a certain level the ECU triggers a cleaning cycle.

Because the diesel engine always operates with excess air, the exhaust gas contains enough oxygen that at temperatures above approximately 550°C, soot will burn off of its own accord. The exhaust gas temperature in a diesel engine is normally between 150°C and 200°C, which is not high enough to burn off the soot. When the cleaning cycle is triggered, an additive is injected into the fuel which increases the exhaust gas temperature and burns off the soot. The additive is stored in a container next to the fuel tank.

If the filter is still clogged following a cleaning cycle, the ECU will go into 'limp home' mode and engine performance will be severely impaired. The filter must then be cleaned by a dealer workshop or diesel specialist using a special diagnostic tool in conjunction with the system's ECU on-board the vehicle. (This cleaning may also be specified as a routine maintenance operation, typically every 48 000 miles.) The engine's high pressure injection system is utilized to inject fuel into the exhaust gases during the post-injection period; this causes the filter temperature to increase sufficiently to oxidize the particulates, leaving an ash residue. The filter is then removed from the exhaust system, and the ash residue flushed away with water.

Exhaust gas recirculation system

An exhaust gas recirculation (EGR) system is designed to recirculate small quantities of exhaust gas into the inlet tract, and therefore into the combustion process. This process reduces the level of oxides of nitrogen present in the final exhaust gas which is released into the atmosphere, and also lowers the combustion temperature.

The volume of exhaust gas recirculated is controlled by vacuum, via a solenoid valve. The solenoid valve is controlled by a fuel injection pump-mounted sensor on models with a conventional injection pump, or by the ECU on models with an electronically-controlled injection system *(see illustration)*.

1 Common rail
   injection system
2 High pressure fuel
   pump
3 Engine
4 Fuel tank
5 Fuel lift pump
6 Additive injector and
   regulator
7 Additive tank
8 Exhaust gas pressure
   sensors
9 Catalytic converter
10 Burn-off filter
11 Silencer
12 Exhaust gases

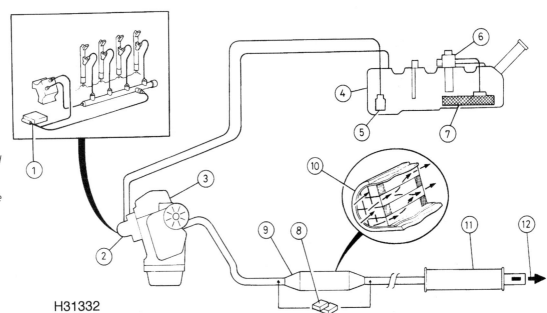

H31332

Schematic view of emission control system using soot burn-off filter

1 Brake servo vacuum
   hose
2 Vacuum converter
   (fitted to fuel
   injection pump)
3 Recirculation valve
   (fitted to exhaust
   manifold)
4 Flow valve/butterfly
   housing (fitted to
   inlet manifold)
5 Electronic control unit
6 Solenoid valve
7 Coolant temperature
   sensor
8 Crankshaft
   speed/position sensor

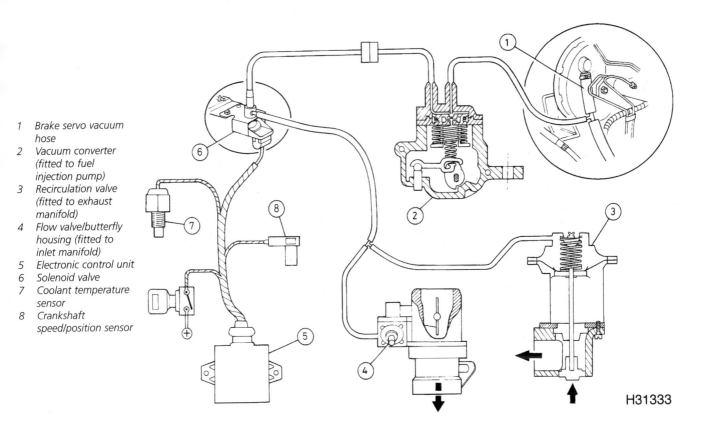

H31333

Schematic view of a typical exhaust gas recirculation system

A vacuum-operated recirculation valve is fitted to the exhaust manifold, to regulate the quantity of exhaust gas recirculated. The valve is operated by the vacuum supplied via the solenoid valve.

Between idle speed and a pre-determined engine load, power is supplied to the solenoid valve, which allows the recirculation valve to open. Under full-load conditions, the exhaust gas recirculation is cut off. On most EGR systems, additional control is provided by the engine temperature sensor, which cuts off the vacuum supply until the engine coolant temperature reaches a pre-determined level, preventing the recirculation valve from opening during the engine warm-up phase.

Knock and smoke

The image of the diesel engine for many years was of a noisy, smoky machine, and to some extent this was justified. It is worth examining the causes of knock and smoke, both to see how they have been reduced in modern engines, and to understand what causes them to get worse.

There is inevitably a small delay (typically 0.001 to 0.002 second) between the start of fuel injection and the beginning of combustion. This delay, known as ignition lag, is greatest when the engine is cold and idling. The characteristic diesel knock is caused by the sudden increase in cylinder pressure which occurs when the injected fuel has been mixed with the hot air and starts burning. It is therefore an unavoidable part of the combustion process, though it has been greatly reduced by improvements in combustion chamber and injection system design. A defective injector (particularly one which is not atomising the fuel as it should for optimum combustion) will also cause the engine to knock.

Smoke is caused by incorrect combustion, but unlike knock it is more or less preventable. During start-up and warm-up a certain amount of white or blue smoke may be seen, but under normal running conditions the exhaust should be clean. The thick black smoke which is all too familiar from old or badly-maintained vehicles is caused by a lack of air for combustion, either because the air intake is restricted (clogged air cleaner), or because too much fuel is being injected (defective injectors or pump).

3 Fuel supply and injection systems

Fuel injection systems and components are covered in detail in Chapter 3. This Section gives an overview of the systems used, and their basic principles of operation.

Fuel supply

The fuel supply system is concerned with delivering clean fuel, free of air, water or other contaminants, to the injection pump. It always includes a fuel filter and a water trap (which may be combined in one unit), a fuel tank, and the associated pipework. Some arrangement must also be made for returning excess fuel from the fuel injectors and the fuel injection pump to the tank (see illustration).

On older vehicles which use an in-line injection pump, or where the fuel tank outlet is significantly lower than the injection pump, a fuel lift pump is used between the tank and the filter. When a distributor injection pump is fitted, and the tank outlet is at about the same level as the injection pump (as is the case with most passenger cars), a separate lift pump is not fitted. In this case, a hand-priming pump is often provided for use when bleeding the fuel system.

Additional refinements may be encountered. These include a fuel heater, which may be integral with the filter, or

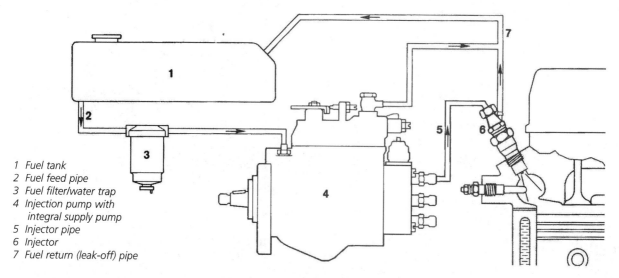

1 Fuel tank
2 Fuel feed pipe
3 Fuel filter/water trap
4 Injection pump with
   integral supply pump
5 Injector pipe
6 Injector
7 Fuel return (leak-off) pipe

Fuel circulation - typical passenger car system
© Robert Bosch Limited

between the tank and the filter, to prevent the formation of wax crystals in the fuel in cold weather. On some vehicles, a 'water-in-fuel' warning light may be illuminated by a device in the water trap when the water reaches a certain level.

The water trap and fuel filter are vital for satisfactory operation of the fuel injection system. On some vehicles, the water trap may have a glass bowl, in which case water build-up can be seen, or it may as already mentioned have some electrical device for alerting the driver to the presence of water. Whether or not these features are present, the trap must be drained at specific intervals, or more frequently if experience shows this to be necessary. If water enters the injection pump it can cause rapid corrosion, especially if the vehicle is left standing for any length of time.

The fuel filter may be of the disposable cartridge type, or it may consist of a renewable element inside a metal bowl (see illustrations). Sometimes a coarser pre-filter is fitted upstream of the main filter. Whatever the type, it must be renewed at the specified intervals. Considering the damage which can be caused to the injection equipment by the entry of even small particles of dirt, it is not worth using cheap replacement filters, which may not be of the same quality as those of reputable manufacture.

Fuel injection pump

In a conventional diesel injection system, the pump is a mechanical device attached to the engine, driven at half-engine speed by a chain, gears or toothed belt. Its function is to supply fuel to the injectors at the correct pressure, at the correct moment in the combustion cycle, and for the length of time necessary to ensure efficient combustion. The pump responds to depression of the accelerator pedal by increasing fuel delivery, within the limits allowed by the

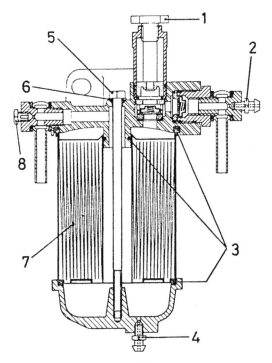

Sectional view of a typical fuel filter

1 Hand-priming plunger
2 Fuel bleed screw (on outlet union)
3 Seals
4 Water drain tap
5 Through bolt
6 Through bolt seal
7 Filter element
8 Air bleed screw (on inlet union)

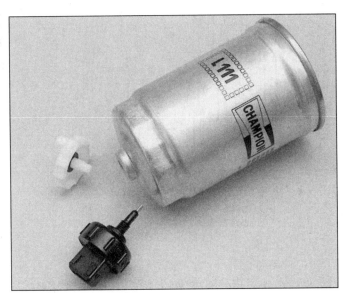

Cartridge type fuel filter showing the drain screw components on the base of the filter

Bosch PE in-line injection pump and associated components
© Robert Bosch Limited

1 Pump
2 Governor housing
3 Lift pump
4 Drivegear and advance mechanism

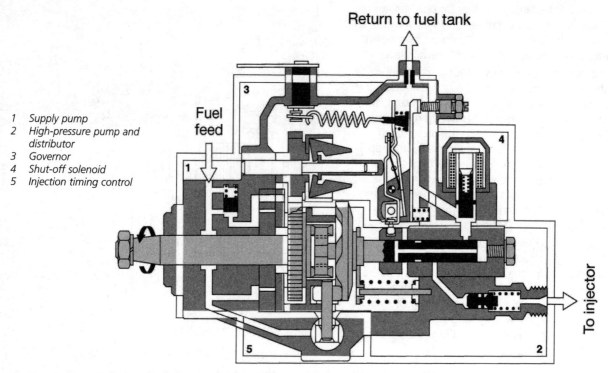

1 Supply pump
2 High-pressure pump and
  distributor
3 Governor
4 Shut-off solenoid
5 Injection timing control

Return to fuel tank

Fuel
feed

To injector

Sectional view of Bosch VE distributor injection pump
© Robert Bosch Limited

governor. It is also provided with some means of cutting off fuel delivery when it is wished to stop the engine.

There are two basic types of pump; the in-line pump, generally fitted to larger engines, and the distributor pump, commonly fitted to passenger car engines. The in-line pump has one pump plunger per engine cylinder. The distributor pump, as its name implies, has a single pump plunger and directs its output to each cylinder in turn (see illustrations).

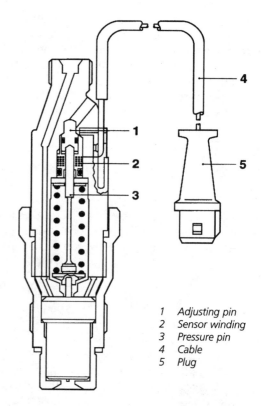

1 Adjusting pin
2 Sensor winding
3 Pressure pin
4 Cable
5 Plug

Fuel injector with needle motion sensor for electronic diesel control

© Robert Bosch Limited

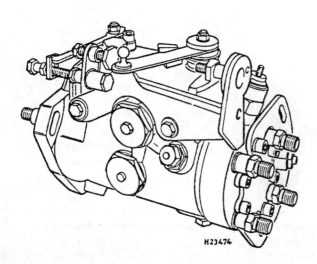

H23474

Lucas/CAV DPC-type distributor injection pump

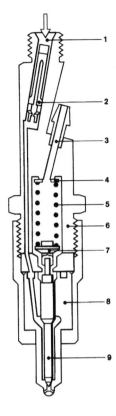

Sectional view of a multi-hole injector
© Robert Bosch Limited

1 Fuel inlet
2 Integral filter
3 Fuel return
4 Pressure adjusting shim
5 Spring
6 Body
7 Spindle
8 Nozzle body
9 Nozzle needle

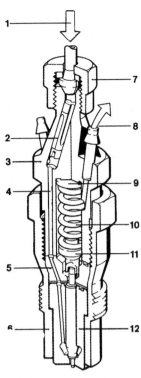

Cutaway view of a pintle injector
© Robert Bosch Limited

1 Fuel inlet
2 Integral filter
3 Body
4 Pressure passage
5 Sleeve
6 Nozzle retainer
7 Union nut
8 Fuel return
9 Pressure adjusting shim
10 Spring
11 Spindle
12 Nozzle

Some kind of governor is associated with the injection pump, either integral with it or attached to it. All vehicle engine governors regulate the fuel delivery to control idle speed and maximum speed; the variable-speed governor also regulates the intermediate speeds. Operation of the governor may be mechanical or hydraulic, or it may be controlled by manifold depression.

Other devices in, or attached to, the pump include cold start injection advance or fast idle units, turbo boost pressure sensors, and anti-stall mechanisms.

Fuel injection pumps are normally very reliable. If they are not damaged by dirt, water or unskilled adjustment, they may well outlast the engine to which they are fitted.

Some modern electronically-controlled diesel injection systems use alternatives to the conventional in-line or distributor fuel pumps – details are given in Section 4.

Fuel injectors

One fuel injector is fitted to each cylinder. The function of the injector is to spray an evenly-atomised quantity of fuel into the combustion or pre-combustion chamber when the fuel pressure exceeds a certain value, and to stop the flow of fuel cleanly when the pressure drops. Atomisation is achieved by a spring-loaded needle which vibrates rapidly against its seat when fuel under pressure passes it. The needle and seat assembly together are known as the injector nozzle.

Injectors in direct injection engines are usually of the multi-hole type, while those in indirect engines are of the pintle type. The 'throttled pintle' injector gives a progressive build-up of injection, which is valuable for achieving smooth combustion (see illustrations).

The injector tips are exposed to the temperatures and pressures of combustion, so not surprisingly they will in time suffer from carbon deposits and ultimately from erosion and burning. Service life will vary according to factors such as fuel quality and operating conditions, but typically one could expect to clean and recalibrate a set of injectors after about 50 000 miles (80 000 km), and perhaps to renew them or have them reconditioned after 100 000 miles (161 000 km).

Some modern electronically-controlled diesel injection systems use electronically-controlled injectors – details are given in Section 4.

Injector pipes

The injector pipes are an important part of the system, and must not be overlooked. The dimensions of the pipes are important, and it should not be assumed that, just because the end fittings are the same, a pipe from a different engine can be used as a replacement. Securing clips must be kept tight, and the engine should not be run without them, as damage from vibration or fuel cavitation may result.

Note that on common-rail engines, all manufacturers recommend replacing the fuel delivery pipes between the pump and the accumulator rail and the injectors once they have been disturbed, as it is possible for minute metal particles to enter them as a result of tightening the union nuts. If these particles enter the fuel injectors, fuel at high-pressure can enter the combustion chambers unrestricted.

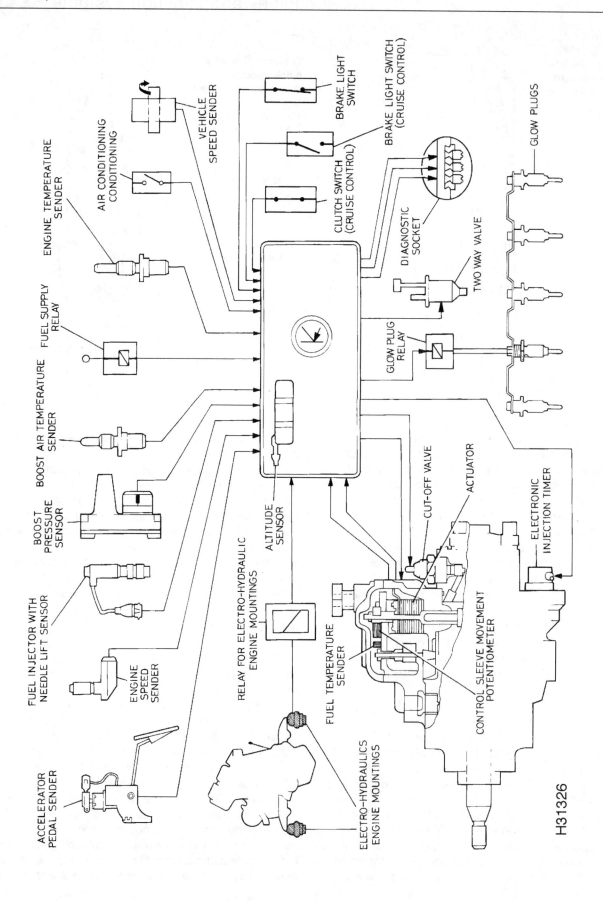

BRAKE LIGHT SWITCH

BRAKE LIGHT SWITCH (CRUISE CONTROL)

GLOW PLUGS

VEHICLE SPEED SENDER

AIR CONDITIONING CONDITIONING

ENGINE TEMPERATURE SENDER

CLUTCH SWITCH (CRUISE CONTROL)

DIAGNOSTIC SOCKET

TWO WAY VALVE

FUEL SUPPLY RELAY

GLOW PLUG RELAY

FUEL TEMPERATURE SENDER

BOOST AIR TEMPERATURE SENDER

BOOST PRESSURE SENSOR

CUT-OFF VALVE

ACTUATOR

ELECTRONIC INJECTION TIMER

FUEL INJECTOR WITH NEEDLE LIFT SENSOR

ENGINE SPEED SENDER

ALTITUDE SENSOR

RELAY FOR ELECTRO-HYDRAULIC ENGINE MOUNTINGS

CONTROL SLEEVE MOVEMENT POTENTIOMETER

ACCELERATOR PEDAL SENDER

ELECTRO-HYDRAULICS ENGINE MOUNTINGS

H31326

Electronic diesel control system components fitted to an Audi 2.5 litre engine

4 Electronic diesel engine control systems

Development of the diesel engine, and particularly the fuel injection system, has been relatively slow compared with the advances which have been made in petrol engine fuel injection and management systems. However, in recent years, electronic diesel engine control systems have been developed to improve diesel engine efficiency and to reduce exhaust emissions. Almost all modern engines use some form of electronic engine control system.

For a diesel engine to operate efficiently, it is essential that the correct amount of fuel is injected at the correct pressure, and at exactly the right time. Even small deviations can cause increased exhaust emissions, increased noise, and increased fuel consumption. In a typical diesel engine, the injection process takes only a thousandth of a second, and only a minute quantity of fuel is injected.

Electronic control using a conventional fuel injection pump

As we've already seen, the function of the fuel injection pump is to supply fuel to the injectors at the correct pressure, at the correct moment in the combustion cycle,

and for the length of time necessary to ensure efficient combustion. A conventional (mechanically-controlled) fuel injection pump uses an accelerator cable (connected to the driver's accelerator pedal), and various mechanical add-on devices (such as cold start injection advance, fast idle units, turbo boost pressure sensors, etc) to provide control of the fuel injection timing and the quantity of fuel injected. Even with these add-on devices, it has become increasingly difficult for a mechanical diesel control system to keep pace with modern demands on engine refinement and exhaust emission control.

Many electronic diesel engine control systems use a conventional in-line or distributor fuel injection pump, but the injection pump timing and the quantity of fuel injected are controlled electronically instead of mechanically. Various electronic sensors are used to measure variables such as accelerator pedal position, engine crankshaft speed, engine camshaft position, the mass of air passing into the engine, turbocharger boost pressure, engine coolant temperature, ambient air temperature, etc *(see illustrations)*.

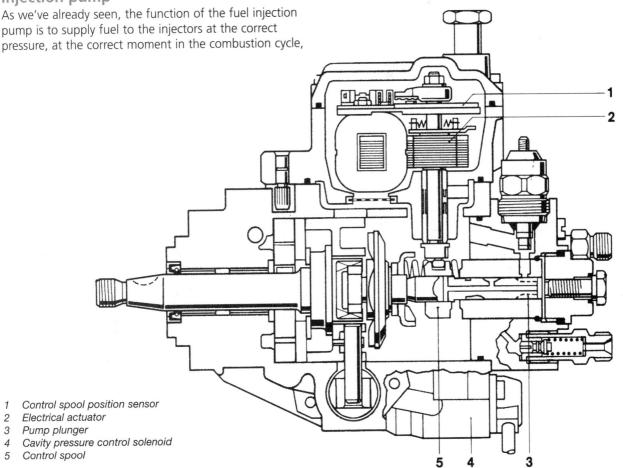

1 *Control spool position sensor*
2 *Electrical actuator*
3 *Pump plunger*
4 *Cavity pressure control solenoid*
5 *Control spool*

Bosch VE injection pump with electronic diesel control
© Robert Bosch Limited

The information from the various sensors is passed to an electronic control unit (ECU), which evaluates the signals. The ECU memory contains a series of mapped values for injected fuel quantity, and start-of-injection point. The ECU performs a number of calculations based on the information provided by the sensors, and selects the most appropriate values for the fuel quantity and start-of-injection point from its stored values. The ECU is capable of analysing the data and performing calculations many times per second, which allows very accurate control over the operation of the engine.

Common rail diesel injection systems

The most widespread common rail system in current use is the Bosch system. Although there are other types of common rail system (eg, Caterpillar system), we will use the Bosch type as a typical example to explain the principles involved.

The common rail system derives its name from the fact that a common rail, or fuel reservoir, is used to supply fuel to all the fuel injectors. Instead of an in-line or distributor fuel pump, which distributes the fuel directly to each injector, a high-pressure pump is used, which generates a very high fuel pressure (up to 1350 bar on some systems) in the accumulator rail. The accumulator rail stores fuel, and maintains a constant fuel pressure, with the aid of a pressure control valve. Each injector is supplied with high-pressure fuel from the accumulator rail, and the injectors are individually controlled via signals from the system electronic control unit. The injectors are electromagnetically-operated.

In addition to the various sensors used on models with a conventional fuel injection pump, common rail systems also have a fuel pressure sensor. The fuel pressure sensor allows the electronic control unit to maintain the required fuel pressure, via the pressure control valve.

For the purposes of describing the operation of a common rail injection system, the components can be divided into three sub-systems: the low-pressure fuel system, the high-pressure fuel system and the electronic control system.

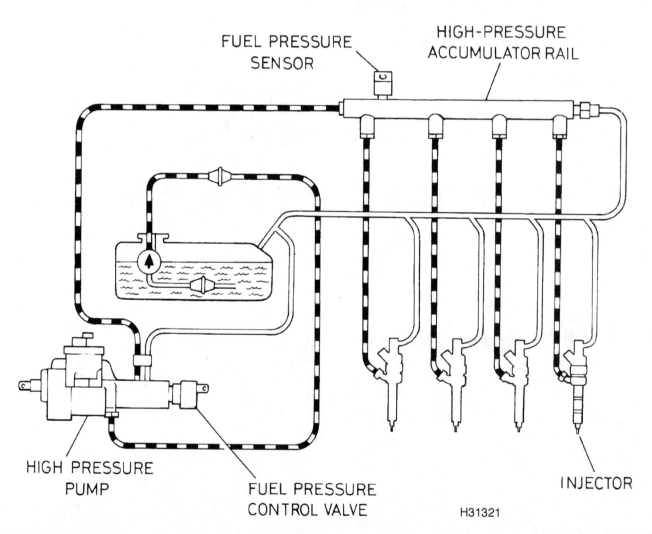

FUEL PRESSURE SENSOR

HIGH-PRESSURE ACCUMULATOR RAIL

HIGH PRESSURE PUMP

FUEL PRESSURE CONTROL VALVE

INJECTOR

H31321

Schematic view of a high-pressure fuel system - Bosch common rail

Low-pressure fuel system

The low-pressure fuel system may consist of the following components:

> Fuel tank.
> Fuel lift pump.
> Fuel filter/water trap.
> Low-pressure fuel lines.
> Fuel cooler

The low-pressure system (fuel supply system) is responsible for supplying clean fuel to the high-pressure fuel circuit.

High-pressure fuel system

The high-pressure fuel system consists of the following components *(see illustration)*:

> High-pressure fuel pump with pressure control valve.
> High-pressure accumulator rail with fuel pressure regulator.
> Fuel injectors.
> High-pressure fuel lines.

After passing through the fuel filter, the fuel reaches the high-pressure pump, which forces it into the accumulator rail, generating pressures of up to 1350 bar. As diesel fuel has a certain elasticity, the pressure in the accumulator rail remains constant, even though fuel leaves the rail each time one of the injectors operates: additionally, a pressure control valve mounted on the high-pressure pump ensures that the fuel pressure is maintained within pre-set limits.

The *pressure control valve* is operated by the ECU. When the valve is opened, fuel is returned from the high-pressure pump to the tank, via the fuel return lines, and the pressure in the accumulator rail falls. To enable the ECU to trigger the pressure control valve correctly, the pressure in the accumulator rail is measured by a *fuel pressure sensor*.

The electromagnetically-controlled fuel injectors are operated individually, via signals from the ECU, and each injector injects fuel directly into the relevant combustion chamber. The fact that high fuel pressure is always available allows very precise and highly flexible injection in comparison to a conventional injection pump: for example combustion during the main injection process can be improved considerably by the pre-injection of a very small quantity of fuel.

Electronic control system

The electronic control system consists typically of the following components:

> Electronic control unit (ECU).
> Fuel lift pump.
> Crankshaft speed/position sensor.
> Camshaft position sensor.
> Accelerator pedal position sensor.
> Turbocharger boost pressure sensor.
> Air temperature sensor.
> Coolant temperature sensor.
> Air mass meter.
> Fuel pressure sensor.
> Fuel injectors.
> Fuel pressure control valve.
> Preheating control circuit.
> EGR valve actuator.

The information from the various sensors is passed to the ECU, which evaluates the signals. The ECU contains electronic 'maps' which enable it to calculate the optimum quantity of fuel to inject, the appropriate start of injection, and even pre- and post injection fuel quantities, for each individual engine cylinder under any given condition of engine operation.

Additionally, the ECU carries out monitoring and self-diagnostic functions. Any faults in the system are stored in the ECU memory, which enables quick and accurate fault diagnosis using appropriate diagnostic equipment (such as a suitable fault code reader).

⚠ Caution

Common rail injection systems operate at higher pressures and finer tolerances than distributor pump systems. They are extremely vulnerable to damage by dirt or water contamination. To avoid premature failure of expensive components, observe the following precautions:

- Take great care not to allow dirt or water to get into fuel lines when working on the system

- If the 'water in fuel' warning light comes on, drain or renew the fuel filter immediately

- Avoid buying fuel from outlets with a low turnover - the longer the fuel has been in storage, the more chance it has of containing water

- Renew components such as fuel lines whenever they are disturbed, when this is specified by the manufacturer

- Don't run out of fuel. This can cause extensive damage: if the fuel lift pump runs dry, it can produce metallic particles which then travel to the high pressure pump and injectors

Chapter 1

Schematic view of the Bosch/VW pump injector system

1 Fuel tank
2 Fuel cooler
3 Fuel temperature sensor
4 Pressure limiting valve
5 Bypass
6 Fuel distributor pipe
7 Pump injectors
8 Cylinder head

9 Restrictor
10 Fuel pump
11 Strainer
12 Pressure limiting valve
13 Non-return valve
14 Fuel filter

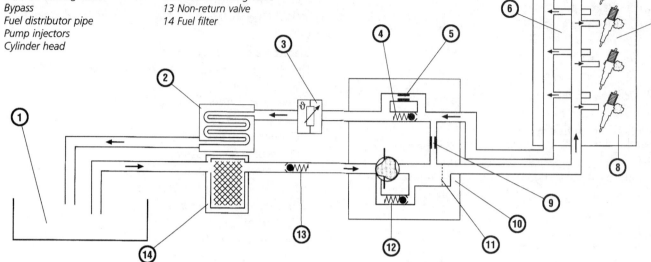

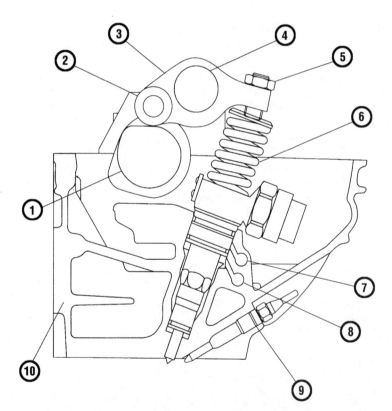

Pump injector installation - Bosch/VW pump injector system

1 Camshaft injection lobe
2 Rocker roller
3 Rocker
4 Rocker shaft
5 Ball-pin adjuster
6 Pump injector
7 Fuel return line
8 Fuel supply line
9 Glow plug
10 Cylinder head

'Pump injector' systems

The 'pump injector' system has been in use in basic form for some years on larger direct injection diesel engines *(see illustration)*. Recent developments in electronic engine control systems have enabled the system to be refined for use on smaller car and light commercial engines, and at the time of writing VW/Audi, and Land Rover were among the major manufacturers selling vehicles equipped with this system. Although there are other types of pump injector system (eg, Lucas EUI), we will use the Bosch type as a typical example to explain the principles involved.

As its name implies, a 'pump injector' consists of a fuel injection pump, combined with a fuel injector. Each cylinder of the engine has its own pump injector, which eliminates the need for a separate high-pressure fuel pump, and the associated high-pressure fuel lines.

The pump injectors are operated by the engine camshaft, and are able to generate extremely high fuel pressures (up to 2000 bar on some systems) *(see illustration)*. The pump injectors are mounted in the cylinder head, and are supplied with fuel via a distributor pipe mounted in the cylinder head. A fuel lift pump pumps fuel from the fuel tank to the distributor pipe. Each pump injector is individually controlled via signals from the system electronic control unit.

The pump injectors are electromagnetically-operated.

Pressure limiting valves maintain constant fuel pressures in the fuel supply and return lines.

Because of the extremely high fuel injection pressure, the fuel in the return line becomes very hot, and a fuel cooling system is used to cool the excess fuel before it is returned to the tank. Besides the obvious effect on safety, if the fuel was not cooled, the fuel temperature in the tank would rise, which means that the temperature of the fuel supplied to the injectors would also rise. Under high-pressure injection conditions, hot fuel reduces fuel delivery from the injectors; although the ECU can compensate to a reasonable degree for fuel temperature variations, cool fuel gives improved combustion and hence improved engine efficiency.

For the purposes of describing the operation of a pump injector system, the components can be divided as follows: the low-pressure fuel system, the fuel cooling system, the pump injectors and the electronic control system.

Low-pressure fuel system

The low-pressure fuel system consists of the following components:

Fuel tank.

Low-pressure fuel lines.

Fuel filter/water trap.

Fuel lift pump (incorporating pressure limiting valve).

Fuel distributor pipe (mounted in cylinder head).

The low-pressure system (fuel supply system) is responsible for supplying clean, cool fuel to the pump injectors.

After passing through the filter, the fuel reaches the fuel lift pump, which supplies fuel to the fuel distributor pipe, via passages drilled in the cylinder head.

Any excess fuel is returned from the distributor pipe to the fuel tank, via the fuel cooling system.

Fuel cooling system

The fuel cooling system is separate from the engine cooling circuit, because the temperature of the engine coolant is too high to cool the fuel when the engine is at operating temperature. In most cases, the fuel coolant circuit is connected to the main coolant expansion tank, but in such a way that the hotter engine coolant circuit has no adverse effect on the fuel coolant circuit. The connection to the expansion tank allows the system to be filled, and also allows for expansion of the coolant with varying temperature.

A fuel cooler may be mounted on the fuel filter head *(see illustration)*. The fuel cooler is basically a fuel/coolant heat exchanger. Cold coolant is pumped through the cooler by an electric pump, controlled by the engine ECU. As the coolant passes through the cooler, it absorbs heat from the fuel. The cooled fuel then passes to the fuel tank, while the warm coolant passes to a radiator at the front of the vehicle. The radiator, which is separate from the engine cooling system radiator, is cooled by the air passing through it due to the forward motion of the vehicle, supplemented by air from the engine cooling fan(s) when necessary. The cold coolant then passes to the coolant pump, and the cycle starts again.

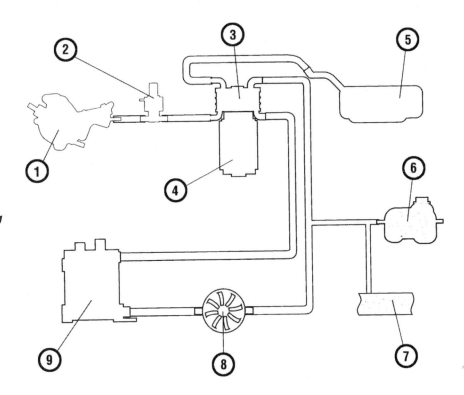

Schematic view of fuel cooling circuit - Bosch/VW pump injector system

1 *Fuel pump*
2 *Fuel temperature sensor*
3 *Fuel cooler*
4 *Fuel filter*
5 *Fuel tank*
6 *Coolant expansion tank*
7 *Engine cooling system*
8 *Electric coolant pump*
9 *Coolant radiator*

Chapter 1

Pump injectors

After passing through the fuel distributor pipe, the fuel reaches the pump injectors.

The electromagnetically-controlled pump injectors are operated individually, via signals from the ECU, and each injector injects fuel directly into the relevant combustion chamber. The fact that very high fuel pressure is always available allows very precise and highly flexible injection in comparison to a conventional injection pump: for example combustion during the main injection process can be improved considerably by the pre-injection of a very small quantity of fuel. On some systems, the individual control of the injectors also allows individual engine cylinders to be 'switched off' during part-load conditions, to improve fuel economy.

Electronic control system

The electronic control system consists typically of the following components:

Electronic control unit (ECU).
Fuel coolant pump.
Crankshaft speed/position sensor.
Camshaft position sensor.
Accelerator pedal position sensor.
Air temperature sensor.
Coolant temperature sensor.
Air mass meter.
Inlet manifold pressure sensor.
Fuel temperature sensor.
Clutch and brake pedal switches.
Fuel injectors.
Preheating control circuit.
EGR valve actuator.

The information from the various sensors is passed to the ECU, which evaluates the signals. The ECU contains electronic 'maps' which enable it to calculate the optimum quantity of fuel to inject, the appropriate start of injection, and even pre- and post injection fuel quantities, for each individual engine cylinder under any given condition of engine operation.

Additionally, the ECU carries out monitoring and self-diagnostic functions. Any faults in the system are stored in the ECU memory, which enables quick and accurate fault diagnosis using appropriate diagnostic equipment (such as a suitable fault code reader).

5   Biodiesel questions and answers

Q	A
Q What is biodiesel?	A Fuel produced from renewable sources - typically vegetable oil, either new or used (eg waste cooking oil).
Q What are its advantages?	A The plants from which the vegetable oil is produced have absorbed carbon dioxide and given off oxygen while they were growing, so the fuel is 'carbon neutral' - it does not contribute to global warming. Also, it's biodegradable, which means that spillages won't pollute so badly.
Q Is it cheaper than ordinary diesel?	A That depends on how it's taxed. It's more expensive to produce than ordinary diesel, but most European governments give it tax breaks to make it the same price, or slightly cheaper, at the pump. It's normally sold as a blend of 95% normal diesel to 5% biodiesel.
Q Can any diesel car use it?	A Check your car's handbook. Some vehicle manufacturers are quite happy with it, while others say it will void the warranty if you use it. All French diesel has had a 5% biodiesel component for a couple of years now and nothing terrible seems to have happened.
Q Where can I buy it?	A The number of filling stations offering biodiesel in the UK is small but growing (150 in mid-2004).
Q Can I make it at home?	A In theory yes, but it's unlikely to be an economical proposition, especially if you cost your own labour. Some of the chemicals involved are pretty unpleasant, and you'll have to pay tax on the end product.
Q Can I just run the car on cooking oil, then?	A Yes, but not for long. Unmodified cooking oil will clog up the filter and the injectors and lead to rapid engine wear. And it's illegal to use on the road unless you've paid tax on it first.

Chapter 1

Routine maintenance and servicing

1   Introduction

Due to the high working pressure, loads and temperatures found in a diesel engine, the recommended service intervals (especially oil change intervals) are generally more frequent thank those for a comparable petrol engine. Frequent oil changes are particularly important for a diesel engine, as dirt or soot builds up in the oil during normal operation, leading to the deterioration of the oil lubricating qualities.

The vehicle manufacturer's service schedule should always be followed, and it is important to use good quality lubricants, which meet the manufacturer's recommendations.

This Chapter does not provide model-specific procedures for maintenance operations, its purpose is to provide a general guide to operations which are particularly important, or unique to diesel engines. Examples of such operations are:

Engine oil and filter renewal
Draining water from the fuel filter/water separator
Fuel filter renewal
Fuel system bleeding
Fuel injection pump checks and adjustments
Fuel injector checks
Timing belt renewal
Turbocharger boost pressure check

Chapter 2

2 Maintenance Schedules

This is a typical maintenance schedule as recommended by the vehicle manufacturer. Servicing intervals are determined by mileage or time elapsed - this is because fluids and systems deteriorate with age as well as with use. Follow the time intervals if the appropriate mileage is not covered within the specified period.

Vehicles operating under adverse conditions may need more frequent maintenance. Adverse conditions include climatic extremes, full-time towing or taxi work, driving on unmade roads, and a high proportion of short journeys. Consult the appropriate Haynes Service and Repair Manuals for model-specific maintenance schedules.

Every 250 miles (400 km), weekly, or before a long journey

☐ Check engine oil level and top up if necessary
☐ Check coolant level and top up if necessary
☐ Check exhaust smoke
☐ Check operation of glow plug warning light

Every 6000 miles (10 000 km) or 6 months, whichever comes first

☐ Renew engine oil and oil filter
Note: *Frequent oil and filter changes are good for the engine. We recommend changing the oil at the mileage specified here, or at least twice a year if the mileage covered is a less.*

Every 12 000 miles (20 000 km) or 12 months, whichever comes first

☐ Check condition and tension of auxiliary drivebelts
☐ Check coolant strength
☐ Check cooling system hoses for condition and security
☐ Check EGR system components
☐ Renew fuel filter
☐ Check fuel system hoses and pipes for condition and security
☐ Check exhaust emissions
☐ Check pressure sensing hoses and vacuum pipes for condition and security

Every 24 000 miles (40 000 km) or 2 years

☐ Check crankcase vent hoses for condition and security
☐ Check engine breather PCV valve
☐ Renew engine coolant
☐ Renew air filter element

Every 48 000 miles (80 000 km) or 4 years

☐ Renew timing belt

3 Component locations

The following are under bonnet component location photographs for a selection of popular models *(see illustrations)*. For detailed removal and refitting procedures, refer to the appropriate Haynes Service and Repair Manual.

Audi A4

1.9 litre

1 Engine oil filler cap
2 Turbocharger wastegate control
3 Fuel injection pump
4 EGR valve
5 Fuel filter
6 Airflow meter
7 Engine management ECU

Chapter 2

Audi 100 & A6

1.9 litre

1 Engine oil filler cap
2 Airflow meter
3 EGR valve
4 Air filter housing
5 Turbocharger
6 Fuel filter

Audi 100 & A6

2.5 litre

1 Engine oil filler cap
2 Fuel injection pump
3 Airflow meter
4 EGR valve
5 Vacuum pump
6 Injection pump drive belt
7 Fuel filter

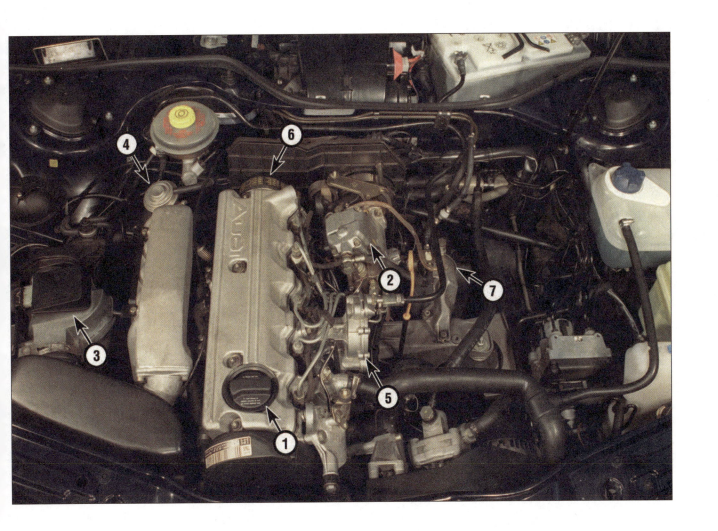

Citroën Saxo

1.5 litre

1 Engine oil filler cap
2 Fuel injection pump
3 Vacuum pump
4 Oil filter
5 Fuel priming 'bulb'
6 Air filter housing
7 Fuel filter

Citroën Xanita

1.9 litre

1 Engine oil filler cap/dipstick
2 Fuel injection pump
3 Fuel priming 'bulb'
4 Engine oil filter
5 Air filter housing
6 Alternator
7 Fuel filter

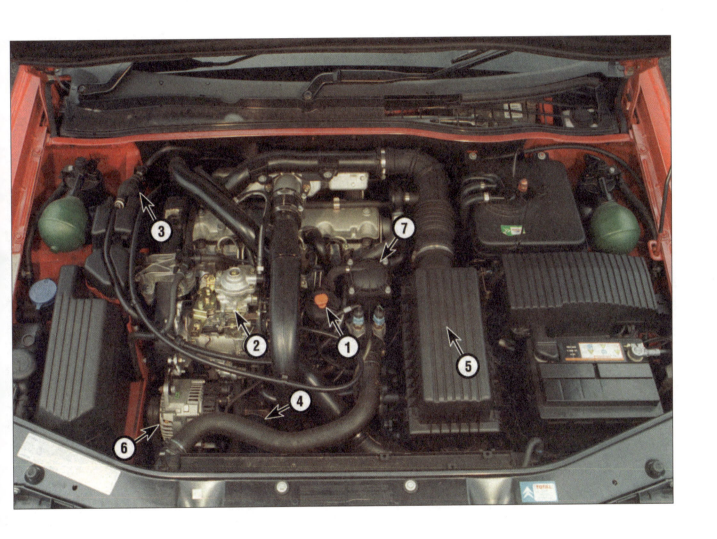

Chapter 2

Citroën Xantia

2.1 litre

1 *Engine oil fillercap/dipstick*
2 *Fuel injection pump*
3 *Fuel priming 'bulb'*
4 *Engine management ECU*
5 *Air filter housing*
6 *EGR pipe*
7 *Fuel filter*

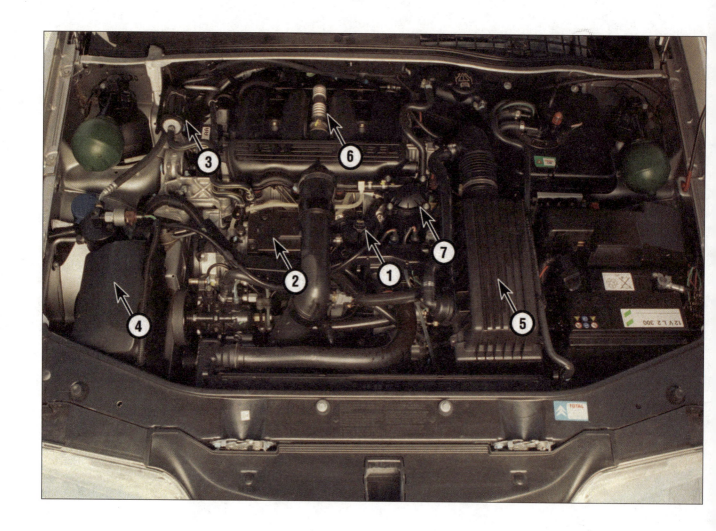

Citroën Xsara

2.0 litre HDI

1 Engine oil filler cap
2 High-pressure fuel pump
3 Accelerator pedal position
  sensor
4 Preheating ECU
5 Airflow meter
6 Engine management ECU
7 Fuel filter

Fiat Punto

1.7 litre Turbo

1 Engine oil filler cap
2 Fuel injection pump
3 Air filter housing
4 Power steering pump
5 Oil filter
6 Fuel filter

Ford Escort & Orion

Non-turbo

1 Engine oil filler cap
2 Fuel injection pump
3 Air filter housing
4 Vacuum pump
5 Idle-up control unit
6 Fusebox
7 Fuel filter

Ford Escort & Orion

Turbo

1 Engine oil filler cap
2 Fuel injection pump
3 Fuel priming pump
4 Air filter housing
5 Intercooler
6 Fuel filter

Ford Fiesta

1.8 litre

1 Engine oil filler cap
2 Fuel injection pump
3 Air filter housing
4 Thermostat housing
5 EGR valve
6 Fuel filter

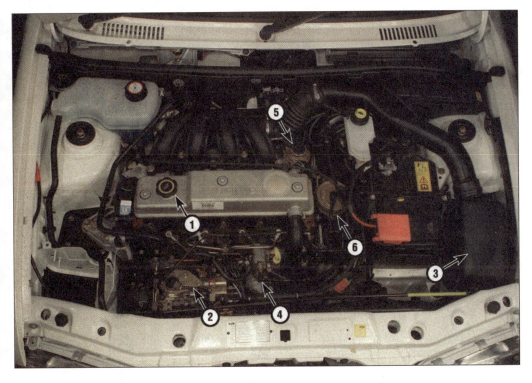

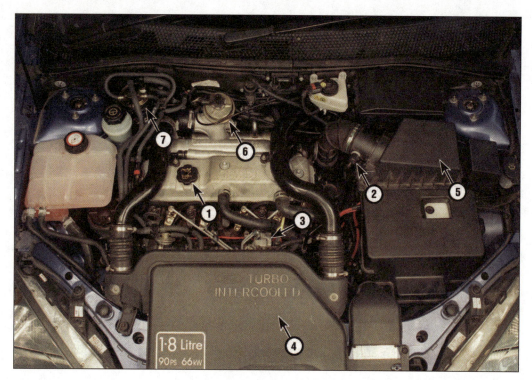

Ford Focus

1.8 litre

1 Engine oil filler cap
2 Airflow meter
3 Fuel injection pump
4 Intercooler
5 Air filter housing
6 EGR valve
7 Fuel filter

Ford Mondeo

2.0 litre

1 Engine oil filler cap
2 Airflow meter
3 Accumulator rail
4 Pressure sensor
5 Turbocharger wastegate
   actuator
6 Air filter housing
7 Fuel filter

Ford Galaxy

1.9 litre

1 Engine oil filler cap
2 Fuel injection pump
3 Air filter housing
4 Airflow meter
5 Intake air temperature sensor
6 Injector with needle lift
   sensor
7 Fuel filter

Land Rover Discovery

300 TDi

1 Engine oil filler cap
2 Fuel injection pump
3 Air filter housing
4 Preheating control unit
5 Engine breather filter
6 Fusebox
7 Fuel filter

Land Rover Freelander

TD4

1 *Engine oil filler cap*
2 *High-pressure fuel pump*
3 *Fuel lift pump*
4 *EGR control valve*
5 *Air cleaner*
6 *Crankshaft position sensor*
7 *Fuel filter*

Land Rover
Freelander

L-Series

1 Engine oil filler cap
2 Needle lift sensor
3 Fuel shut-off solenoid
4 Fuel injection pump
5 Engine management ECU
6 Turbocharger
7 Fuel filter

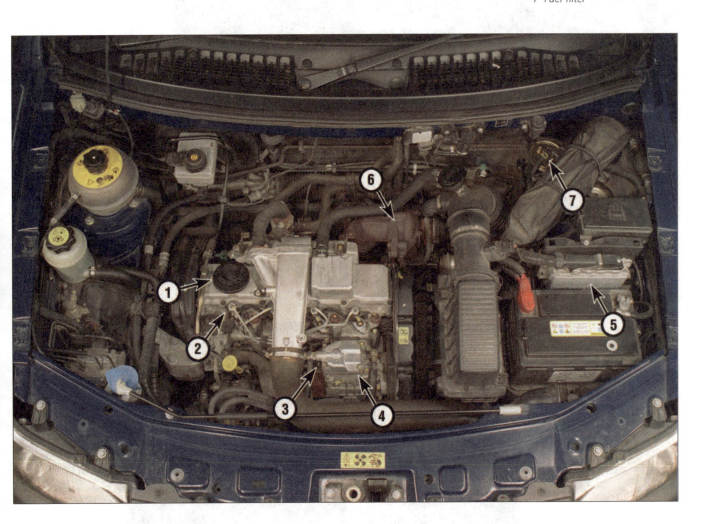

Chapter 2

Mercedes 190

2.5 litre

1 Engine oil filler cap
2 Oil filter
3 In-line fuel injection pump
4 Air filter housing
5 Fusebox
6 Fuel filter

Mercedes C-Class

2.5 litre

1 Engine oil filler cap
2 Oil filter
3 EGR valve
4 Diagnostic socket
5 Air filter housing
6 Fuel injection pump
7 Fuel filter

Peugeot 206

1.9 litre

1 Engine oil filler cap
2 Air filter housing
3 Vacuum pump
4 Preheating ECU
5 Engine management ECU
6 Fuse/relay box
7 Fuel filter

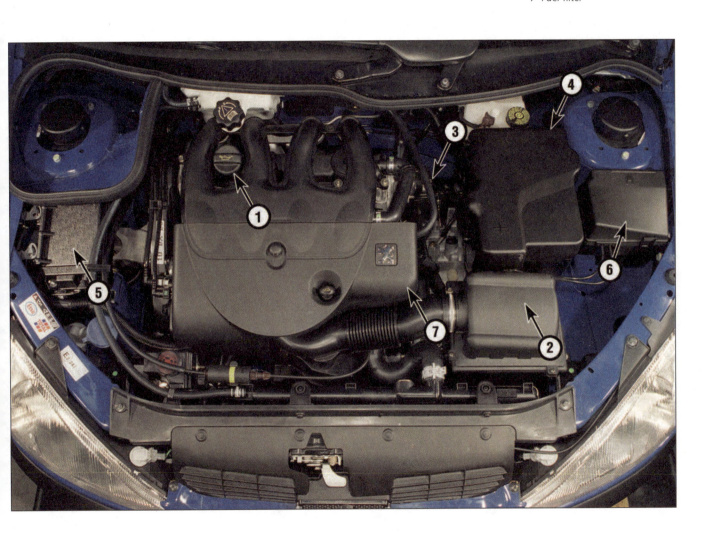

Peugeot 206

2.0 litre HDi

1 Engine oil filler cap
2 Engine management ECU
3 Airflow meter
4 Vacuum pump
5 Fuel inertia shut-off valve
6 Fuse/relay box
7 Fuel filter

Peugeot 306

1.9 litre Turbo model

1 Engine oil filler cap/dipstick
2 Fuel injection pump
3 Fuel priming 'bulb'
4 Intercooler
5 Oil filter
6 Fuse/relay box
7 Fuel filter

Chapter 2

Peugeot 306

1.9 litre Non-turbo model

1 *Engine oil filler cap/dipstick*
2 *Fuel injection pump*
3 *Fuel priming 'bulb'*
4 *Oil filter*
5 *Vacuum pump*
6 *Fuse/relay box*
7 *Fuel filter*

Peugeot 405

Turbo model

1 Engine oil filler cap/dipstick
2 Fuel injection pump
3 Air filter housing
4 Intercooler
5 Fuel priming 'bulb'
6 Fuel filter

Peugeot 405

Non-turbo model

1 Engine oil filler cap/dipstick
2 Fuel injection pump
3 Fuel priming 'bulb'
4 Vacuum pump
5 Air filter housing
6 Fuel shut-off valve
7 Fuel filter

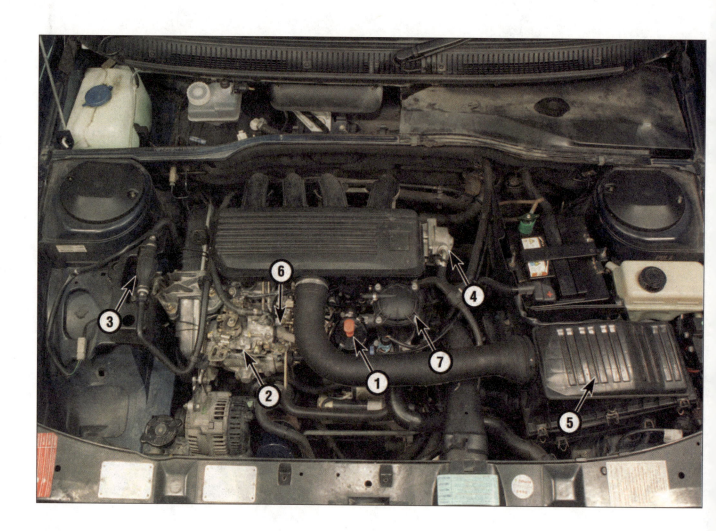

Peugeot 406

2.1 litre

1 Engine oil filler cap/dipstick
2 Fuel injection pump
3 Fuel priming 'bulb'
4 Accelerator pedal position
   sensor
5 Fuel inertia cut-off valve
6 Vacuum pump
7 Fuel filter

Peugeot 406

2.2 litre

1 Engine oil filler cap
2 Airflow meter
3 Fuel injection pump
4 Hydraulic pump
5 Air filter housing
6 Throttle pedal position sensor
7 Fuel filter

Renault Clio

'91 - '96

1 Engine oil filler cap
2 Fuel injection pump
3 Fuel shut-off valve solenoid
4 Fuel priming pump
5 Vacuum pump
6 Air filter housing
7 Fuel filter

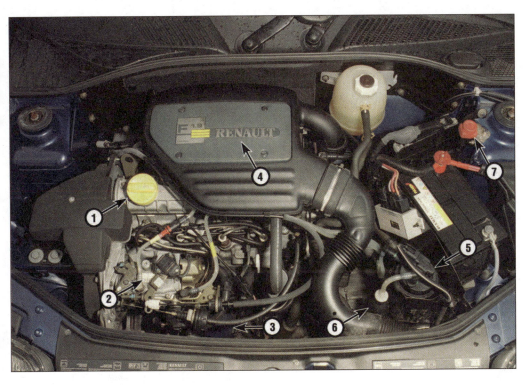

Renault Clio

'98 - '01

1 Engine oil filler cap
2 Fuel injection pump
3 Engine oil filter
4 Air filter housing
5 Fuel filter
6 Fuel priming 'bulb'
7 Fuel inertia cut-off valve

Renault Espace

2.1 litre

1 Engine oil filler cap
2 Fuel injection pump
3 Turbocharger
4 Fuel priming pump
5 Thermostat housing
6 Air filter housing
7 Fuel filter

Renault Laguna

1.9 litre

1 Engine oil filler cap
2 Fuel injection pump
3 Airflow meter
4 Accelerator pedal position
   sensor
5 EGR valve
6 Fuel priming 'bulb'
7 Fuel filter

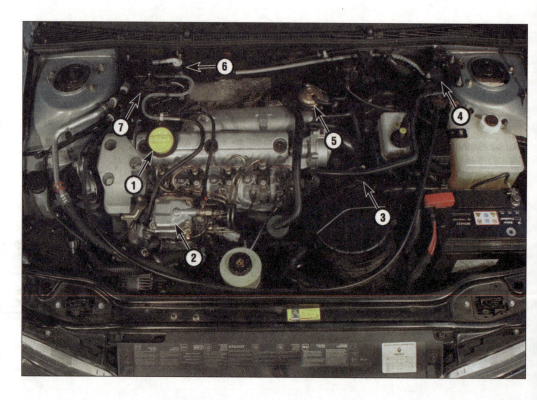

Renault Laguna

2.2 litre

1 Engine oil filler cap
2 Fuel injection pump
3 EGR valve
4 Fuel priming pump
5 Air filter housing
6 Inlet manifold
7 Fuel filter

Renault Megane

'96 - '98

1 Engine oil filler cap
2 Fuel injection pump
3 Fuel priming 'bulb'
4 Air filter housing
5 Cold start capsule
6 Vacuum pump
7 Fuel filter

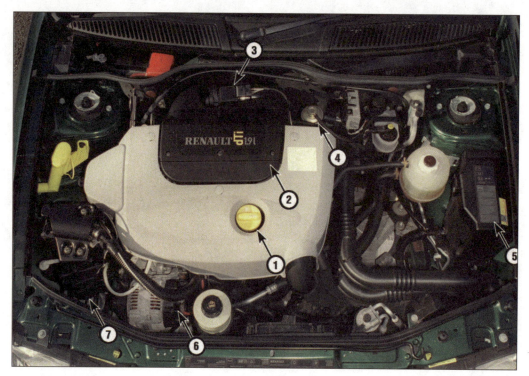

Renault Megane

'99 - '02

1 Engine oil filler cap
2 Air filter housing
3 Airflow meter
4 EGR control valve
5 Fuse/relay box
6 Alternator
7 Fuel filter

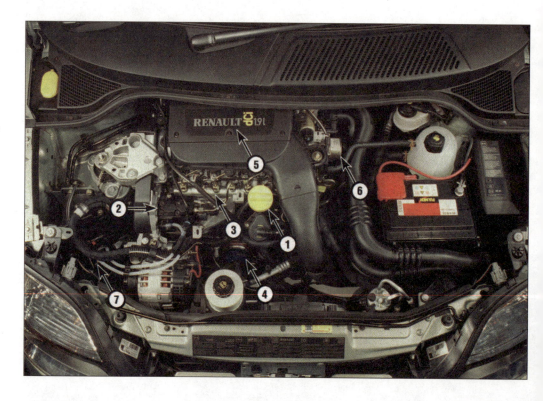

Renault Scenic

'99 - '02

1 Engine oil filler cap
2 High-pressure fuel pump
3 Accumulator rail
4 Engine oil filter
5 Air cleaner housing
6 Vacuum pump
7 Fuel filter

Rover 420 D

2.0 litre

1 Engine oil filler cap
2 Airflow meter
3 Engine management ECU
4 Air filter housing
5 Fuel priming 'bulb'
6 Alternator with integral
   vacuum pump
7 Fuel filter

Seat Ibiza & Cordoba

1.9 litre

1 Engine oil filler cap
2 Fuel injection pump
3 Airflow meter
4 Air temperature sensor
5 Fuel shut-off valve
6 Turbo boost pressure sensor
7 Fuel filter

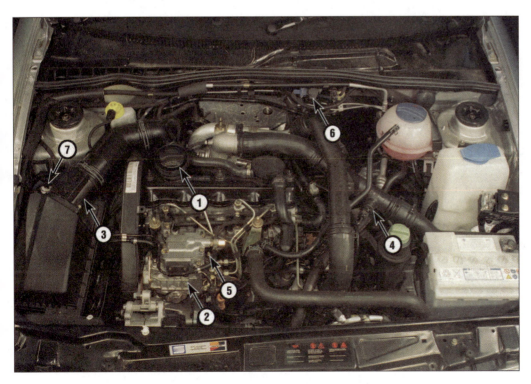

Skoda Felicia

1.9 litre

1 Engine oil filler cap
2 Fuel injection pump
3 EGR valve
4 Air filter housing
5 Preheating control unit
6 Idle speed boost valve
7 Fuel filter

Vauxhall Astra/Zafira

1.7 litre DOHC

1 Engine oil filler cap
2 Air filter housing
3 Airflow meter
4 Engine management ECU
5 Fuse/relay box
6 Fuel injection pump
7 Fuel filter

Vauxhall
Astra/Zafira

1.7 litre SOHC

1 Engine oil filler cap
2 Fuel injection pump
3 Airflow meter
4 EGR valve
5 Air filter housing
6 Fuse/relay box
7 Fuel filter

Vauxhall
Frontera

2.3 litre

1 Engine oil filler cap
2 Fuel injection pump
3 Air filter housing
4 Thermostat housing
5 Intercooler pipe
6 Fuel filter and priming pump

Vauxhall Frontera

2.5 litre

1 Engine oil filler cap
2 Airflow meter
3 Air filter housing
4 Fuel priming pump
5 Alternator with integral vacuum pump
6 Fuel filter

Vauxhall Frontera

2.8 litre

1 Engine oil filler cap
2 Fuel priming pump
3 Air filter housing
4 Throttle valve actuator
5 Thermostat housing
6 Turbocharger
7 Fuel filter

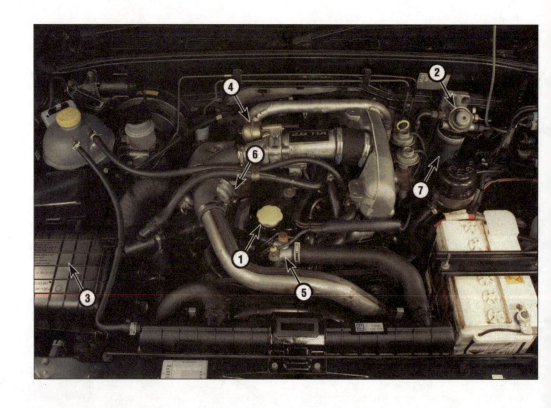

Vauxhall Vectra

2.0 litre

1 Engine oil filler cap
2 Turbocharger
3 Vacuum pump
4 Air filter housing
5 Airflow meter
6 Fuse/relay box
7 Fuel filter

VW Golf/Bora

1.9 litre

1 Engine oil filler cap
2 Fuel injection pump
3 Airflow meter
4 Engine oil filter
5 EGR valve
6 Vacuum pump
7 Fuel filter

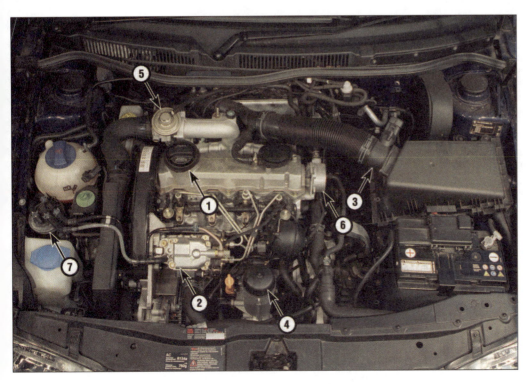

VW Passat

1.9 litre TDi

1 Engine oil filler cap
2 Airflow meter
3 Inlet air temperature sensor
4 Fuel injection pump
5 Air filter housing
6 Vacuum pump
7 Fuel filter

VW Polo

1.9 litre

1 Engine oil filler cap
2 Fuel injection pump
3 Air filter housing
4 EGR valve
5 Engine coolant temperature
   sensor
6 Fuel filter

4 Lubricants and fluids

Modern diesel engines place great demands upon the lubricants and fluids used in their operation. The use of high technology in lubricant and fluid refinement has resulted in servicing intervals being continuously extended. Many engine manufacturers now specify engine oil change intervals of up to 20 000 miles, and coolant systems filled 'for life'. The drive for longer intervals benefits manufacturers and operators of large vehicle fleets - it does not necessarily benefit the private owner.

We recommend changing the oil at least once a year, regardless of the mileage covered. This way the oil will always be in good condition, and hopefully the engine internals as well.

Although perhaps not as critical as engine oil, the coolant may also benefit from being changed every four years or so. Corrosion of the cooling system components and blockages of the internal passages of the water jacket are the major causes of overheating in all engines. Note also, that whenever an aluminium part which has contact with the coolant is changed, the coolant must be changed at the same time, to replenish the coolant's resistance to corrosion.

5 Maintenance procedures

Note: *This Section does not provide an exhaustive list of diesel engine maintenance procedures, it gives basic information and advice on tasks which are especially important or unique to diesel engines. Always refer to the manufacturer's information for a detailed description of maintenance operations.*

Engine oil and filter renewal

Frequent oil and filter changes are the most important preventative maintenance procedures which can be undertaken by the DIY owner. As engine oil ages, it becomes diluted and contaminated, which leads to premature engine wear.

Before starting this procedure, gather together all the necessary tools and materials. Also make sure that you have plenty of clean rags and newspapers handy, to mop up any spills. Ideally, the engine oil should be warm, as it will drain better, and more built-up sludge will be removed with it. Take care, however, not to touch the exhaust or any other hot parts of the engine when working under the vehicle. To avoid any possibility of scalding, and to protect yourself from possible skin irritants and other harmful contaminants in used engine oils, it is advisable to wear gloves when carrying out this work. Access to the underside of the vehicle will be greatly improved if it can be raised on a lift, driven onto ramps, or jacked up and supported on axle stands. Whichever method is chosen, make sure that the vehicle remains level, or if it is at an angle, that the drain plug is at the lowest point. Where fitted, release the screws/nuts/clips and remove the engine undershield.

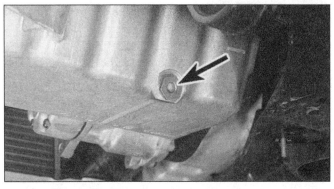

Undo the engine sump drain plug (arrowed)

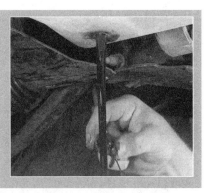

(!) *As the drain plug releases from the threads, move it away sharply so the stream of oil issuing from the sump runs into the container, not up your sleeve*

Slacken the drain plug about half a turn, position the draining container under the drain plug, then remove the plug completely *(see illustration)*. If possible, try to keep the plug pressed into the sump while unscrewing it by hand the last couple of turns. Recover the sealing ring from the drain plug.

Allow some time for the old oil to drain, noting that it may be necessary to reposition the container as the oil flow slows to a trickle.

After all the oil has drained, wipe off the drain plug with a clean rag, and fit a new sealing washer where applicable *(see illustration)*. Clean the area around the drain plug opening, and refit the plug. Tighten the plug to the correct torque where specified (see Chapter 4).

If the filter is also to be renewed, move the container into position under the oil filter, which is located on the front side of the cylinder block.

Renew the sump plug sealing washer (arrowed) - where fitted

Chapter 2

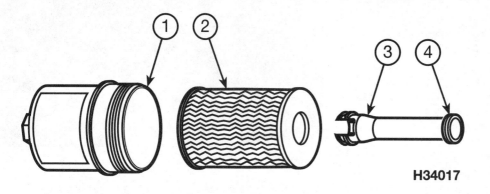

Paper element type filters - typical contents

1 Oil filter cover
2 Oil filter element
3 Plunger tube
4 Plunger tube O-ring

H34017

Undo the oil filter cover (arrowed) - paper element type

Fit a new O-ring to the cover

Paper element filters

On some engines, the paper filter element is contained within a filter cover. Using a socket or spanner, slacken and remove the filter cover *(see illustration)*. Be prepared for fluid spillage, and recover the O-ring seal from the cover.

Pull the filter element from the filter housing.

Use a clean rag to remove all oil, dirt and sludge from the inside and outside of the filter cover.

Fit the new O-ring to the filter cover, then insert the new filter element into the housing. On some models, a locating peg is incorporated into the base of the element. Ensure that the peg engages correctly with the corresponding hole in the housing *(see illustrations)*.

Apply a little clean engine oil to the O-ring seal, then refit the filter/cover to the housing and tighten the cover to the specified torque.

Ensure the filter locating peg (arrowed) locates into the corresponding hole in the housing (arrowed)

2•36

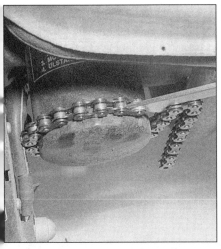

Slacken the canister using a chain wrench ...

.. a strap wrench ...

... or a special tool (arrowed)

Spin-on canister filters

These are one-piece filter canisters which screw onto a threaded sleeve.

Using an oil filter removal tool if necessary, slacken the filter initially, then unscrew it by hand the rest of the way *(see illustrations)*. Empty the oil in the old filter into the container.

Use a clean rag to remove all oil, dirt and sludge from the filter sealing area on the engine. Check the old filter to make sure that the rubber sealing ring hasn't stuck to the engine. If it has, carefully remove it.

Apply a light coating of clean engine oil to the sealing ring on the new filter, then screw it into position on the engine *(see illustration)*. Tighten the filter firmly by hand only – **do not** use any tools. Where necessary, refit the engine undershield.

Apply a little clean engine oil to the sealing ring, and fit the new filter

All engines

Remove the dipstick, then unscrew the oil filler cap. Fill the engine, using the correct grade and type of oil. An oil can spout or funnel may help to reduce spillage. Pour in half the specified quantity of oil first, then wait a few minutes for the oil to fall to the sump. Continue adding oil a small quantity at a time until the level is up to the upper mark on the dipstick *(see illustration)*. Refit the filler cap.

Start the engine and run it for a few minutes; check for leaks around the oil filter seal and the sump drain plug. Note that there may be a delay of a few seconds before the oil pressure warning light goes out when the engine is first started, as the oil circulates through the engine oil galleries and the new oil filter (where fitted) before the pressure builds-up.

Switch off the engine, and wait a few minutes for the oil to settle in the sump once more. With the new oil circulated and the filter completely full, recheck the level on the dipstick, and add more oil as necessary.

Dispose of the used engine oil safely.

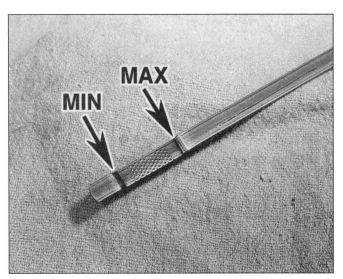

Note the maximum and minimum marks (arrowed) on the dipstick

Chapter 2

Draining water from the fuel filter/water trap

If a glass bowl water trap is fitted, it is easy to see when water is accumulating in the bottom of it. When the water trap is incorporated into the filter base, the water cannot be seen. Sometimes a sensor is fitted, which illuminates a warning light to alert the driver to the presence of water (see illustrations).

Manufacturer's recommendations for the intervals at which the water trap should be drained vary widely. Obviously, operating conditions and fuel quality will determine the rate at which water accumulates, but it is better to err on the side of safety and drain the trap frequently. If water gets through to the pump and injectors, it can cause serious damage.

When draining the water trap, place a small container under it to catch the fuel. It is important that fuel is not allowed to spill onto the coolant hoses, alternator, starter motor or engine mountings. Protect them with plastic sheet if necessary. On some models, the trap or filter is awkwardly placed; in such cases, it may be easier to fit a length of hose to the trap outlet (see illustration).

When the drain screw is opened, it may be found that no fuel emerges because the system is under negative pressure. Slacken the bleed screw or the inlet union on the filter head, or operate the hand-priming pump, until fuel flows (see illustration).

When clean fuel, free of water droplets, flows out, tighten the drain screw and the bleed screw.

Dispose of the drained fuel and water safely, in the same way as used engine oil.

Fuel filter water trap drain plug (arrowed)

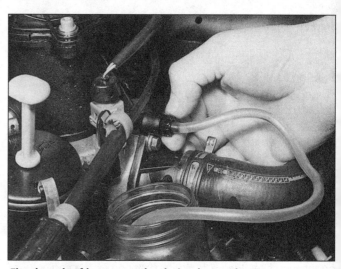

Fit a length of hose over the drain plug outlet

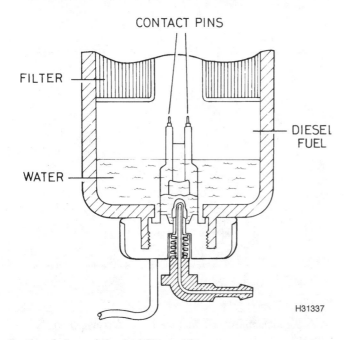

Sectional view of the fuel filter with water sensor

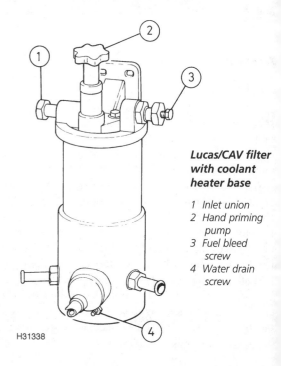

Lucas/CAV filter with coolant heater base

1 Inlet union
2 Hand priming pump
3 Fuel bleed screw
4 Water drain screw

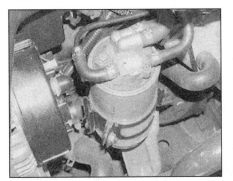

Canister fuel filter fitted to VAG engines ...

... cartridge fuel filter (Peugeot 206) ...

... and Citroën Xantia

Fuel filter renewal

The main filter element must be renewed at the intervals specified by the manufacturer, or more often if experience of particular problems show it to be necessary. Some manufacturers specify renewal at the beginning of every winter, even if little mileage has been covered, to reduce the risk of waxing problems.

Filters are basically of two types: cartridge and canister *(see illustrations)*. Cartridge filters can be subdivided into 'spin-on' type, similar to a modern engine oil filter, 'clamp' type retained by a clamping strap or band, and 'through-bolt' type, retained by a bolt running from the filter head to a separate bowl. Canister filters are totally enclosed in the filter bowl.

It is best to drain the filter before removal if possible. The filter is then unscrewed with a strap or chain wrench ('spin-on' type), or the through-bolt or clamp bolt removed, according to type. Make sure that the old seals are recovered: some filter heads have a seal in a groove which is easy to overlook. The through-bolt, where fitted, may have an O-ring seal under its head. Any imperfect seals can allow air to be drawn into the system if there is no lift pump, or fuel to be forced out if there is *(see illustrations)*.

Wipe out the filter bowl or canister, if applicable, finishing off with a **clean** non-fluffy cloth, or (if available) compressed air.

Caution: It is important that no dirt is introduced into the system.

Smear the new seals with a little clean fuel. If a central seal retainer is fitted, make sure it is secure; in the case of the canister filter, make sure the seal is snugly in its groove.

Fit and secure the new filter or element, then bleed the fuel system if necessary.

If a separate water trap or pre-filter is fitted, this may incorporate a gauze screen, which should be removed for cleaning at the specified intervals.

Undo the through-bolt...

... and remove the cartridge

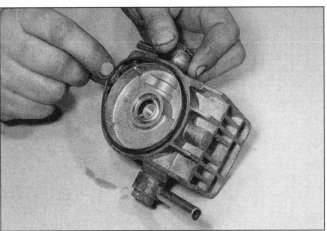

Recover the seal from the filter housing

Fuel system bleeding

Bleeding of the fuel system is necessary after operations in which air has been allowed to enter fuel lines, and after running out of fuel. General procedures are given here: refer to manufacturers information or to the relevant Haynes *Service and Repair Manual* for specific details.

Modern fuel systems are of the self-bleeding type. If no hand-priming pump is fitted, the normal way of bleeding such a system is by cranking the engine on the starter motor in 10-second bursts. If a hand-operated vacuum pump is available, this can be connected to the injection pump fuel return connection and used to suck fuel through the supply lines and filter; this will obviously save the battery a good deal of work.

When a hand-priming pump is fitted, this is operated first, with the bleed screw (where fitted) on the filter head open. When fuel free from air bubbles emerges, tighten the bleed screw. Carry on pumping until increased resistance is felt. Alternatively, use a vacuum pump as just described; this avoids any risk of splitting the diaphragm on the hand-priming pump, an occurrence which is not unknown on older vehicles. Note that on certain models (E.g. Peugeot 307, Ford Fusion), a hand priming pump is fitted, but no bleed screw. On these models, simply operate the pump until fuel free from bubbles appears in the transparent fuel supply pipe, or resistance is felt *(see illustrations)*.

If air has reached the injection pump, this may be bled out at a specific bleed screw if fitted, or (more usually) at the fuel return union.

On engines fitted with an in-line or distributor injection pump, if air has entered the injector pipes, slacken the injector unions, and crank the engine on the starter motor. When fuel emerges, tighten the unions and mop up spilt fuel.

On engine fitted with a common rail injection system, if air has entered the injector pipes, operate the starter is short bursts until the engine starts. Do **not** slacken the injector pipe/accumulator rail unions - once slackened the pipes must be renewed.

When a separate fuel lift pump is fitted, this usually has a hand-priming lever for use when bleeding *(see illustration)*. If the engine has stopped with the lift pump operating arm on top of its cam, it will be necessary to turn the engine before the hand-priming lever can be used.

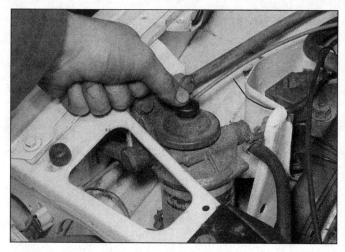

Some priming pumps are integral with the filter housing ...

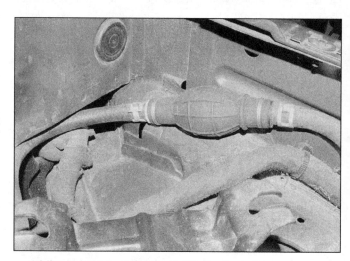

... whilst some are rubber 'bulbs' in the fuel pipe

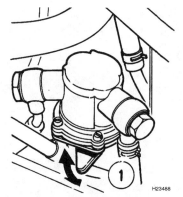

On some lift pumps a hand priming lever (1) is fitted

H23488

Auxiliary drivebelt checks

On some diesel engines, besides the usual auxiliary drivebelt(s) used to drive ancillary units such as the alternator, power steering pump, etc, an additional drivebelt may be used to drive the brake vacuum pump, and in some cases the injection pump (most conventional injection pumps are driven by the engine timing belt).

Where applicable, checking of the vacuum pump and/or injection pump drivebelts should not be overlooked when carrying out routine maintenance. Always renew a drivebelt if there is any doubt about its condition.

Using a suitable socket and bar fitted to the crankshaft pulley bolt, rotate the crankshaft so that the entire length of the drivebelt can be examined. Examine the drivebelt for cracks, splitting, fraying or damage. Check also for signs of glazing (shiny patches) and for separation of the belt plies. Renew the belt if worn or damaged (see illustration).

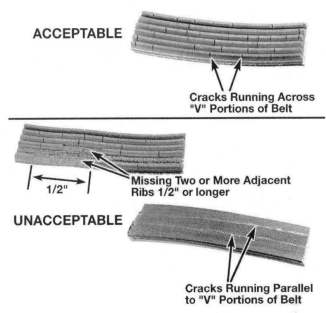

Check the auxiliary belt(s) for signs of wear like this - if it looks worn, replace it

Fuel injection pump checks and adjustments

On some older models, the idle and maximum speeds should be checked at the manufacturer's specified intervals (see Section 6 and refer to the appropriate Service and Repair Manual for procedure details). On later models with EDC (Electronic Diesel Control) or common rail systems, no adjustments are possible, as these functions are under the control of the engine management ECU.

Depending on the type of injection pump and the control systems fitted to it, there may also be a need to check the operation of the anti-stall and cold start devices. Again, procedures are given in the appropriate Service and Repair Manual.

Inspect the injection pump control linkages (where fitted) at every service interval. Lubricate them if necessary, and renew any frayed or sticking cables. Check that fully depressing the accelerator pedal produces full movement of the pump control lever.

At the same intervals, inspect the fuel injector high-pressure pipes and their securing clips for security and condition. Also inspect the fuel return pipes or hoses, and (when applicable) the turbo boost pressure hose which connects the inlet manifold to the injection pump. Renew any leaking or damaged components.

Fuel injector checks

Some manufacturers specify that the injectors should be removed and inspected periodically, but generally they are ignored unless particular problems (excessive smoke, knocking or power loss) suggest that they may be giving trouble.

If the operation of the injectors is suspect, remove them and have them inspected by a main dealer or diesel specialist. Do not attempt to dismantle the injectors.

Fuel injector cleaners are available in the form of fuel additives. If used as directed they are unlikely to be harmful, and may indeed do some good; note however that some vehicle manufacturers specifically forbid their use.

Exhaust emissions check

The only emission test applicable to diesel engines is the measuring of exhaust smoke density. The test involves the use of special test equipment, and forms part of the MoT test for vehicles in the UK.

The test involves accelerating the engine several times to its maximum unloaded speed, and so it is vital to ensure that the engine timing belt is in good condition before the test is carried out. Refer to Chapter 5 for details of possible causes of excessive smoke.

Chapter 2

On most engines, the timing belt drives the injection pump (arrowed) as well as the camshaft ...

... whilst on some, a separate belt drives the pump (Rover L-Series engine shown)

Timing belt renewal

As with petrol engines, if a timing belt is fitted, it is vital to ensure that it is in good condition. On many diesel engines, the timing belt drives the injection pump as well as the camshaft *(see illustrations)*.

The timing belt must be renewed at the manufacturer's specified intervals, or more frequently if the vehicle is used in particularly arduous conditions (e.g., frequent stop-start driving or taxi work).

It is strongly recommended that consideration is given to renewing the timing belt every 48 000 miles (80 000 km), regardless of the manufacturer's recommended renewal intervals.

If the engine leaks oil in the area adjacent to the timing belt, rectify the leak and replace the belt at the earliest opportunity. The oil will cause the condition of the belt to deteriorate rapidly *(see illustrations)*.

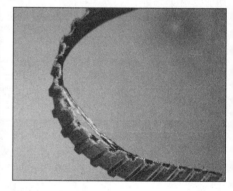

Foreign Body Entrapment

Cause *A foreign body (nut, bolt, washer, etc) has become trapped in the drive and has over-stretched and broken the tensile cords.*

Symptom *Belt breakage, in a curved or ragged tear.*

Remedy *Attempt to locate and identify foreign body*
Ensure belt covers are effective.

Land Wear

Cause *Excessive tension, causing the belt to wear on the pulley lands. Rough sprocket(s) abrading the belt.*

Symptom *Wear, or polishing, on the lands between the teeth, possibly wearing down to the tension cords; with polishing on the tooth crests of trapezoidal belts.*

Remedy *Replace sprocket(s) if required.*

Edge Wear

Cause *Damaged sprocket flange, or misaligned sprockets.*

Symptom *Excessive wear and damage to the belt edges.*

Remedy *Replace damaged sprockets and ensure correct belt alignment.*

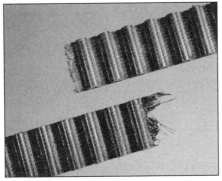

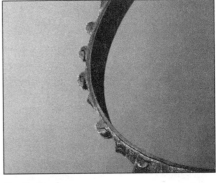

Back Cracks

Cause The rubber has been over-heated and has degraded, possibly from friction on a siezed idler or water pump. Extreme cold may have the same effect.

Symptom A series of cracks across the back of the rubber stock.

Remedy Ensure all spindles driven off the back of the belt, including water pumps, rotate freely.

Tensile Failure

Cause Some of the tensile cord's fibres have broken due to crimping (folding) before or during assembly, creating a weak point.

A belt running over-tensioned may sometimes cause teeth to ride up onto sprocket lands, resulting in vast over-stretching and tensile failure.

Symptom Tensile breakage, with a straight break between two teeth.

Remedy Replace belt carefully, without pinching or levering.

Set new belt to correct tension.

Tooth Peel

Cause Very low tension allowing the belt to jump teeth.

Symptom Teeth peeling, emanating from root cracks. Often is present together with tooth shear.

Remedy Set new belt to correct tension and ensure tensioner mechanism is tight.

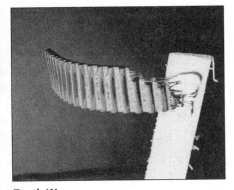

Tooth Wear

Cause Extremely low tension allows the belt to ride out on the sprocket, causing localised wear on edge of the thrust face.

Sometimes excessive tension, pulling the belt up the land, may wear the tooth face, before a tensile failure.

Symptom Hollows through the facing fabric.

Remedy Set new belt to correct tension.

Tooth Shear

Cause May be due to sudden overload of the drive from the seizure of a driven pump, such as a water pump.

Also may be due to low tension, which allows the belt to ride high on the sprocket, producing excessive bending moments, and deflection of the teeth until cracks form.

Symptom Six or more teeth missing, often with cracking in roots of a number of teeth.

Remedy Ensure all driven items rotate freely.

Set new belt to correct tension and ensure tensioner mechanism is tight.

Oil Contamination

Cause Contamination from a failed oil seal, or an oil or diesel leak, breaks down the adhesion of the rubber. Swelling can also cause mis-meshing leading to other types of failure.

Symptom Dirty or smelly belt, with a ragged decomposing structure.

Remedy Ensure oil leak is stopped. Check belt covers and dust shields.

6   Adjustments and checks

On most conventional fuel injection pumps (i.e., pumps without electronic control, non-common rail), it is normally possible to adjust the following settings:

a) *Idle speed.*
b) *Anti-stall controls.*
c) *External controls (e.g., cold idle mechanisms).*
d) *Maximum no-load speed.*
e) *Injection timing.*

Routine adjustments to injection pumps are normally confined to idle speed, anti-stall and external controls, which may include cold idle mechanisms. Checking injection timing is not as routine an operation as checking the ignition timing on a petrol engine. It is necessary when investigating complaints of poor performance, knock and smoke, and whenever the pump or its drive has been disturbed. This last case often includes timing belt renewal.

Some manufacturers also specify a routine check of maximum no-load speed. The screw which controls this speed is always tamperproofed in production, using a locking wire and seal, paint or a sealing cap. *Breaking or removing this tamperproof device may invalidate any manufacturer's warranty.*

Other adjustment screws may be externally accessible, either directly, or through access plugs; they control functions such as maximum-fuelling, excess-fuelling and transfer pressure. Often these screws or plugs are also tamperproofed. Do not attempt haphazard adjustment of such screws. Normally a pump test bench is needed to set (or reset) them correctly.

The following Sections give general procedures. Details specific to particular models may vary; consult manufacturer's information or the appropriate Haynes *Service and Repair Manual* for further information.

Engine speed adjustments

Idle speed

Bring the engine to normal operating temperature, and connect a tachometer to it. (For details of tachometers, see Chapter 6.) If a fast idle device is fitted, make sure that it is not holding the pump control lever or idling lever off its stop.

Allow the engine to idle, and check the speed against that specified in the Service and Repair Manual.

If adjustment is necessary, slacken the locknut and turn the idle speed adjusting screw until the speed is correct. If some tolerance is allowed, adjust the speed to the value within the specified range where the engine runs most smoothly. Tighten the locknut when adjustment is correct (*see illustrations*).

On some pumps, it is necessary to check the anti-stall adjustment if the idle speed is altered.

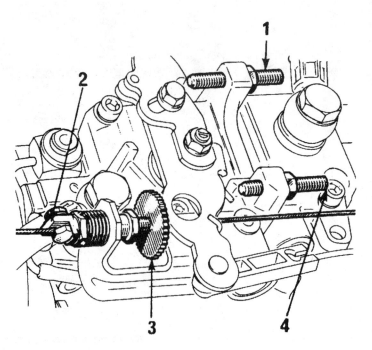

Typical injection pump adjustment points - Bosch VE pump

1   Idle speed adjustment screw
2   Fast idle cable and stop (when fitted)
3   Fast idle adjustment screw (when fitted)
4   Maximum speed adjustment screw

Typical injection pump adjustment points - Lucas/CAV DPC pump

1 Idle speed adjustment screw
2 Locknut
3 Idle speed lever
4 Locknut
5 Anti-stall adjustment screw
6 Maximum speed adjustment screw
X Anti-stall adjustment dimension

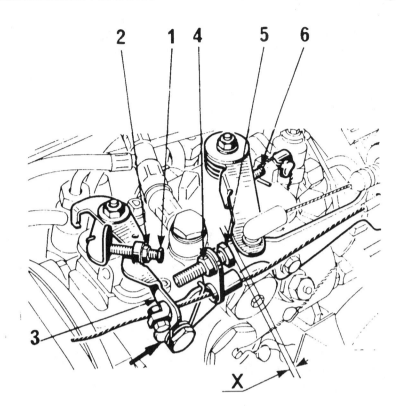

Typical injection pump adjustment points - Bosch VE pump with separate idle lever

1 Fast idle speed adjustment screw
2 Fast idle cable end stop
3 Idle lever
4 Idle speed adjustment screw
5 Anti-stall (residual capacity) adjustment screw
6 Fast idle screw adjuster
7 Accelerator cable adjuster
8 Maximum speed adjustment screw
9 Control lever
a Shim for anti-stall adjustment

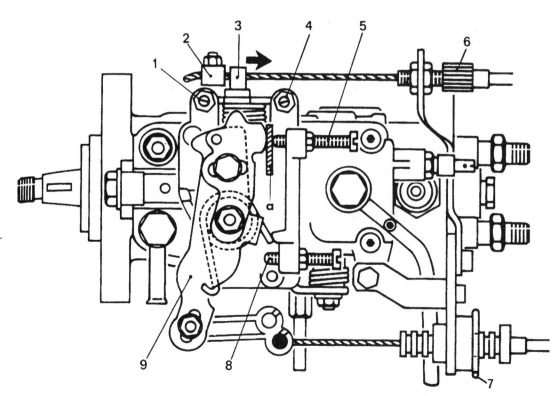

Maximum speed

With the engine warmed up and a tachometer connected, accelerate the engine to maximum speed for a few seconds. Note the speed reached, and compare it with that specified in the Service and Repair Manual. Do not hold maximum speed for any longer than is necessary.

If adjustment is necessary, remove the tamperproofing, slacken the locknut, and turn the adjustment screw. Repeat the check; tighten the locknut and fit a new tamperproof device when adjustment is correct.

Anti-stall (residual capacity)

Anti-stall or residual capacity adjustment determines how quickly engine speed falls off when the accelerator is suddenly released. If the adjustment is incorrect, the engine will either tend to stall after sudden deceleration, or it will 'hang' (fail to lose its speed quickly enough).

All CAV DP series pumps have some kind of external anti-stall adjustment facility, but most Bosch VE pumps do not (see illustration).

When the anti-stall adjustment screw determines the

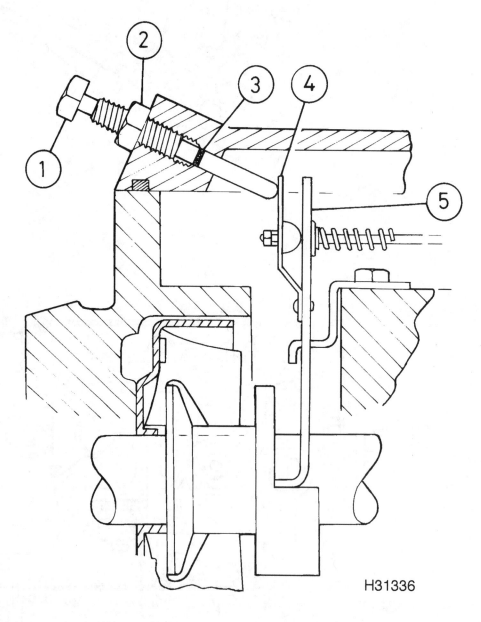

H31336

Separate anti-stall adjustment screw (1) on Lucas/CAV DPC pump with all-speed mechanical governor

2 Locknut   3 Seal   4 Spring   5 Governor arm

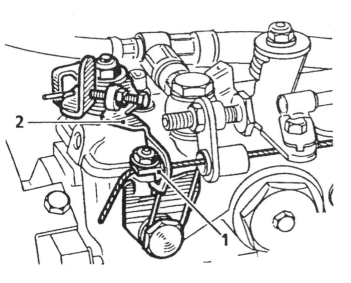

Cold idle adjustment - Lucas/CAV DPC pump with remote thermostatic capsule
When cold, adjust the cable clamp (1) to hold the idle lever (2) against the stop

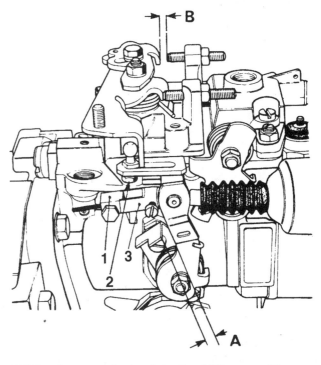

Cold idle adjustment check points - Bosch VE pump with thermostatic capsule

1 Cable end stop
2 Ball-pin adjusting nut
3 Clevis

A Advance lever gap
B Fast idle gap

resting position of the pump control lever, the adjustment procedure usually consists of inserting a specified thickness of shim (or feeler gauge) between the screw and the lever, and noting the effect on idle speed. Idle speed and anti-stall adjustments are connected; if one is adjusted, it will normally be necessary to check the other.

When the anti-stall screw is separate, adjustment is normally on a trial-and-error basis, moving the screw by a quarter-of-a-turn at a time. Turning the screw inwards will reduce the tendency to stall; turning it outwards will reduce the tendency to 'hang'. The effect of a change in adjustment is judged by accelerating the engine to maximum no-load speed, and then releasing the accelerator. The engine must return to idle speed within a specified number of seconds (typically 5 seconds) without stalling.

Cold-idle mechanisms

Cold-idle mechanisms are automatically operated. When in operation, they may affect injection timing, idle speed or both.

Automatic mechanisms which rely on the movement of a lever by a thermostatic capsule and a cable may also require the cable to be adjusted. If the mechanism alters both timing and idle speed, the relationship between the two functions must also be checked (*see illustrations*).

Other types of automatic cold-idle mechanism alter the injection pump timing by raising the transfer pressure. Typically, this is done by an electrically-heated thermostatic capsule, which opens a valve as it warms up; no adjustment is possible.

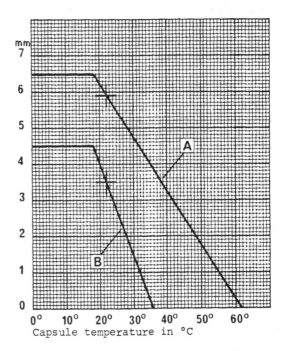

The relationship between the advance lever gap (A) and the fast idle gap (B) varies with temperature - Bosch VE pump with thermostatic capsule

Typical spill tube

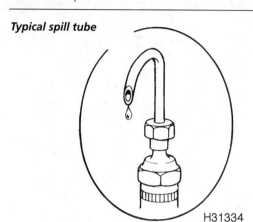

H31334

Pipe connector and delivery valve components

1 Connector
2 O-ring
3 Spring
4 Plunger
5 Sealing washer
6 Valve carrier

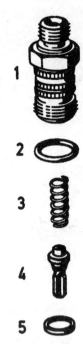

Injection timing

Static timing

Static timing is still the most widely-used method of setting diesel injection pumps. Precision measuring instruments are often needed for dealing with distributor pumps. Good results depend on the skill and patience of the operator.

Caution: Be careful not to introduce dirt into the injection pump during the following procedures.

Spill timing (in-line pumps) – low-pressure method

This is a simple method of timing, albeit messy. The only special equipment required is a 'spill' or 'swan-neck' tube; this can be made in the workshop using part of an old injector pipe if wished *(see illustration)*.

The procedure finds the point in the pump cycle when one plunger covers its inlet port. This corresponds to the beginning of the retraction stroke. It can be accurately related to the engine cycle, but does not necessarily correspond to the actual beginning of injection.

Bring the engine to TDC, No 1 piston firing. Clean around the injector pipe unions and the connections on the pump, then remove No 1 cylinder injector pipe.

Unscrew No 1 cylinder pipe connection from the pump. Remove the delivery valve plunger and spring, noting which way up the plunger is fitted *(see illustration)*. **Do not** remove the valve carrier. Refit the connector and sealing washer, and fit the spill tube to the holder.

Make sure that the stop control is in the 'run' position. In the case of a vacuum-operated stop control, disconnect the vacuum hose. Fix the pump control lever in the maximum-speed position.

The fuel in the pump must now be subject to a small head of fuel pressure. On systems where the fuel filter is higher than the pump, the filter will serve as a header tank. Open the filter bleed screw, and operate the hand-priming pump until fuel emerges.

On systems where the fuel filter is lower than the pump, the necessary pressure can be provided by disconnecting the fuel inlet from the pump, and substituting a feed from a small reservoir of **clean** fuel positioned higher than the pump.

Turn the engine approximately a quarter-turn backwards. Fuel will begin to flow from the tube. Slowly, turn the engine forwards again towards TDC, until the flow of fuel is reduced to one drop per second (or as specified). This is the spill timing point. Note the crankshaft position (degrees BTDC, or alignment of a peg hole), and compare it with that specified.

If adjustment is necessary, slacken the remaining injector pipe unions and the pump mountings. Turn the pump as necessary to advance or retard the timing, then tighten the mountings and repeat the check.

Disconnect the spill tube. Reassemble the delivery valve, using new sealing washers if necessary, being careful not to introduce dirt into the pump. Refit No 1 cylinder injector pipe, and remake the original fuel supply connections if they were disturbed.

Spill timing (in-line pumps) – high-pressure method

If suitable equipment is available, the fuel in the injection pump can be pressurised sufficiently to pass the delivery valve. There is thus no need to dismantle the delivery valve, with a consequent saving in time, although it will be necessary to block off the pump fuel return.

Because of the higher pressure involved, fuel flow from the spill tube will be much faster. The spill timing point is typically defined as the point where the jet of fuel from the spill tube turns into a chain of drops.

Apart from the points just noted, the procedure is the same as for the low-pressure method.

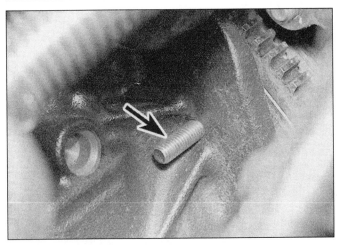

Insert the locking pin (arrowed) through the crankcase into corresponding hole in the flywheel (Peugeot 206) ...

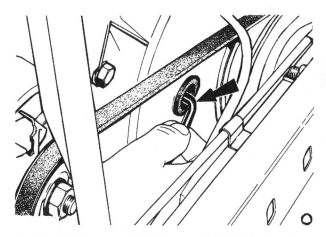

... or through the access hole (arrowed) in the timing cover (Ford 2.5 litre DI) ...

Peg systems (all pump types, when applicable)

Determine the location of the peg holes for the injection pump and the crankshaft or flywheel (as applicable), and the sizes of peg required.

Turn the engine until the crankshaft or flywheel timing peg can be inserted cleanly. With this peg in position, it must be possible to insert the injection pump peg *(see illustrations)*.

CAV DP side-entry

Bring the engine to TDC, No 1 piston firing. Remove the access plug from the side of the injection pump – be prepared for fuel spillage.

Fit a dial test indicator (DTI) and probe so that the probe enters the access hole, passes through the hole in the circlip, and rests on the pump rotor. Slowly turn the crankshaft anti-clockwise to find the DTI minimum reading. In this position, the probe is resting in the bottom of the timing groove in the rotor *(see illustration)*.

Turn the crankshaft clockwise to bring the engine to the specified timing point. This may be TDC, or it may be a specified point before or after TDC – see the appropriate Service and Repair Manual. (If the timing point is overshot, return to the zero position established in the previous paragraph, and start again.)

Read the probe movement displayed on the DTI, and compare it with the specified value.

CAV DP top-entry

Bring the engine to TDC, No 1 piston firing. Remove the access plug from the top of the injection pump.

Insert a probe of the specified length into the access plug hole, so that the tip of the probe rests on the rotor timing piece. Position a dial test indicator to read the movement of the probe.

Turn the crankshaft approximately a quarter-turn backwards, and zero the DTI.

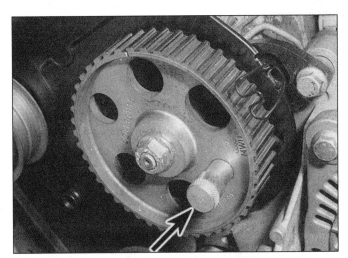

... then insert the pump sprocket locking pin (arrowed)

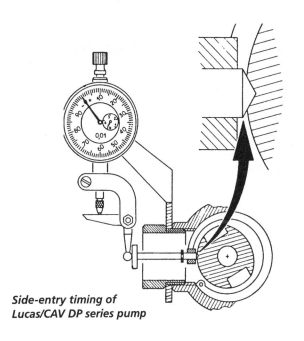

Side-entry timing of Lucas/CAV DP series pump

Reading the probe movement - Lucas/CAV top-entry timing

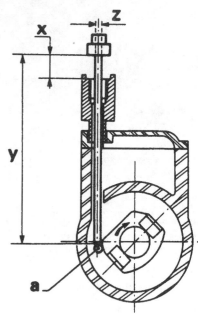

Sectional view of the probe and pump - Lucas/CAV top-entry timing

a *Timing piece*
x *Specified timing value*
y *95.5 ± 0.01 mm*
z *7 mm diameter*

Turn the crankshaft clockwise to bring the engine to the specified timing point (see above). Read the probe movement displayed on the DTI, and compare it with the value engraved on the plastic disc or tag somewhere on the pump *(see illustrations)*.

Bosch VE rear-entry

Bring the engine to TDC, No 1 piston firing. Remove the access plug from the rear of the injection pump – be prepared for fuel spillage *(see illustration)*.

Fit a dial test indicator (DTI), adapter and probe, so that the probe enters the access hole and the DTI displays movement of the pump plunger. Removing the injector pipes will improve access. On some pumps it may be necessary to use a right-angle adapter to allow the DTI to fit in the available space at the rear of the pump *(see illustrations)*.

Slowly turn the crankshaft anti-clockwise until the DTI reading reaches a minimum (pump plunger BDC), and zero the DTI at this point.

Turn the crankshaft clockwise to bring the engine to the specified timing point (see above). Read the plunger movement displayed on the DTI, and compare it with the specified value.

Adjustment (all pump types)

If adjustment is necessary, this may either be carried out by slackening the injector pipe unions and the pump mountings, and turning the pump, or by altering the relationship of the pump drive flange to its gear or sprocket. Refer to the manufacturer's information or to the relevant

Removing the access plug from the rear of a Bosch VE pump

Dial test indicator and adapter positioned with the probe reading the plunger movement - Bosch VE pump rear-entry timing

Haynes Service and Repair Manual for the appropriate method. Repeat the timing check from the beginning after adjustment.

Dynamic timing

As the name implies, dynamic timing is carried out with the engine running. Special equipment is required to carry out dynamic timing accurately, and this is unlikely to be available to the home mechanic. The equipment works by converting pressure pulses in an injector pipe into electrical signals. If such equipment is available, use it in accordance with its manufacturer's instructions.

Although the pump timing is checked with the engine running, any adjustment is usually carried out with the engine stopped.

An additional problem is that few manufacturers specify figures for dynamic timing.

For these reasons, static timing is generally an easier method of timing than dynamic timing, and we have chosen not to include specific details of dynamic timing in this book.

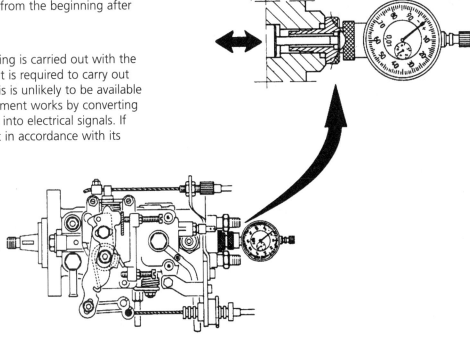

Rear entry timing of a Bosch VE pump

Chapter 2

System components and replacement guidelines

1 Precautions

When working on fuel system components, scrupulous cleanliness must be observed, and care must be taken not to introduce any foreign matter into fuel lines or components.

After carrying out any work involving disconnection of fuel lines, it is advisable to check the connections for leaks; pressurise the system by cranking the engine several times.

Electronic control units are very sensitive components, and certain precautions must be taken to avoid damage to these units as follows.

When carrying out welding operations on the vehicle using electric welding equipment, the battery and alternator should be disconnected.

Although the underbonnet-mounted modules will tolerate normal underbonnet conditions, they can be adversely affected by excess heat or moisture. If using welding equipment or pressure-washing equipment in the vicinity of an electronic module, take care not to direct heat, or jets of water or steam, at the module. If this cannot be avoided, remove the module from the vehicle, and protect its wiring plug with a plastic bag.

Before disconnecting any wiring, or removing components, always ensure that the ignition is switched off.

Do not attempt to improvise fault diagnosis procedures using a test lamp or multi-meter, as irreparable damage could be caused to the module.

After working on fuel injection/engine management system components, ensure that all wiring is correctly reconnected before reconnecting the battery or switching on the ignition.

(!) *Warning: When working on any part of the fuel system, avoid direct contact skin contact with diesel fuel - wear protective clothing and gloves when handling fuel system components. Ensure that the work area is well ventilated to prevent the build up of diesel fuel vapour. Fuel injectors operate at extremely high pressures and the jet of fuel produced at the nozzle is capable of piercing skin, with potentially fatal results. When working with pressurised injectors, take care to avoid exposing any part of the body to the fuel spray. It is recommended that a diesel fuel systems specialist should carry out any pressure testing of the fuel system components.*

Under no circumstances should diesel fuel be allowed to come into contact with coolant hoses - wipe off accidental spillage immediately. Hoses that have been contaminated with fuel for an extended period should be renewed. Diesel fuel systems are particularly sensitive to contamination from dirt, air and water. Pay particular attention to cleanliness when working on any part of the fuel system, to prevent the ingress of dirt. Thoroughly clean the area around fuel unions before disconnecting them. Store dismantled components in sealed containers to prevent contamination and the formation of condensation. Only use lint-free cloths and clean fuel for component cleansing.

2   Fuel system –
priming and bleeding

After any operation which requires the disconnection of any fuel feel pipe/hose, it may be necessary to prime and bleed the fuel system. On many later engines, the fuel systems are self bleeding, simply requiring the starter to be operated until the fuel arrives at the injectors and the engine starts. On other engines, it's necessary to operate a priming pump, whilst a bleed screw on the pump or filter housing is slackened to allow the trapped air to escape. The priming pump is hand operated, and usually takes the form of a rubber 'bulb' in the fuel supply hose, or a simple pump built into the fuel filter housing *(see illustrations)*.

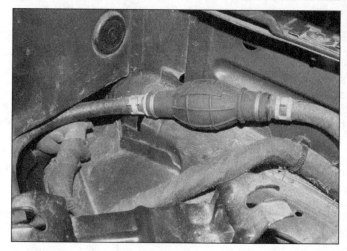

The priming pump is hand operated by squeezing the rubber 'bulb' ...

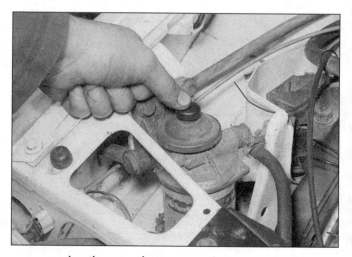

... or pressing the pump button

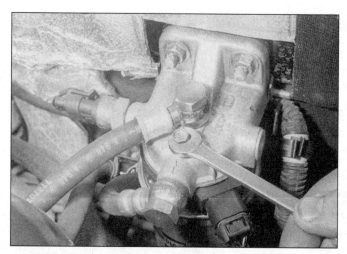

Slacken the bleed screw on top of the fuel filter housing

Locate the bleed screw on the filter housing, or injection pump *(see illustration)*. Undo the screw a turn or so, whilst repeatedly squeezing the bulb, or pressing the pump operating button. On some models, there is no bleed screw - simply operate the priming pump until resistance is felt, or air bubbles no longer appear in the fuel hose. Tighten the bleed screw (where applicable) as soon as fuel free of air begins to emerge.

Depress the accelerator pedal to the floor then start the engine as normal (this may take longer than usual, especially if the fuel system has been allowed to run dry - operate the starter in ten second bursts with 5 seconds rest in between each operation). Run the engine at a fast idle speed for a minute or so to purge any remaining trapped air from the fuel lines. After this time the engine should idle smoothly at a constant speed.

If the engine idles roughly, then there is still some air trapped in the fuel system. Increase the engine speed again for another minute or so then recheck the idle speed. Repeat this procedure as necessary until the engine is idling smoothly.

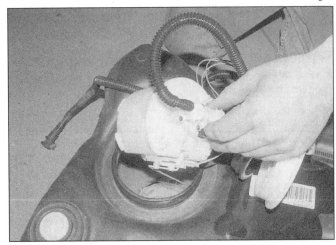

The fuel lift pump may be integral with the fuel gauge sender unit in the tank ...

3   Fuel lift pump

Description

The fuel lift pump supplies fuel at low pressure to the injection pump. Not all models have one; they are not normally fitted to engines which use distributor injection pumps. Originally the lift pump was mechanical and mounted on the engine, where it was driven by a cam; nowadays electrically-operated pumps are widely used, although VAG's pump injector engines use a camshaft-driven pump.

The electrically-operated lift pump may be integral with the fuel gauge sender unit located in the fuel tank, or it may be located elsewhere in the fuel supply circuit *(see illustrations)*. The tank-mounted pump may not be available separately from the fuel gauge sender unit - check with your dealer or fuel injection specialist.

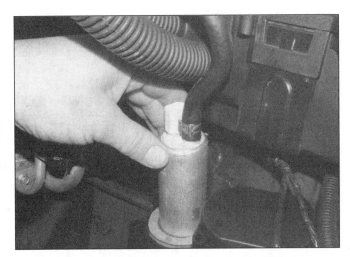

... or be a separate unit (Land Rover Freelander TD4 shown)

Replacement guidelines

Note: *For a detailed step-by-step model-specific removal and replacement procedure, refer to the appropriate Service and Repair Manual.*

❏ Label the fuel pipes attached to the sender unit cover to aid refitment.
❏ Always renew the sender unit cover seal *(see illustration)*.
❏ Apply a thin layer of petroleum jelly to the seal prior to refitting the cover.
❏ Try to remove the pump unit when the tank is almost empty.

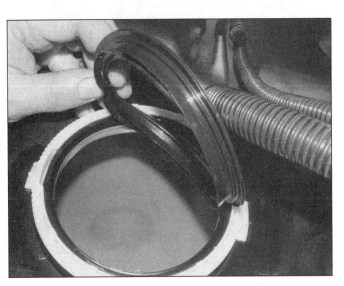

On tank mounted pumps, always renew the sealing ring

Chapter 3

4   Fuel injection pumps

Description

Several designs of injection pump are commonly fitted to modern diesel engines.

In-line and distribution pumps

For traditional indirect and direct injection engines, an in-line or distribution pump is fitted *(see illustrations)*. The function of these pumps is to supply fuel to the injectors at the correct pressure, at the correct moment in the combustion cycle and for the length of time necessary to ensure efficient combustion. The pump responds to depression of the accelerator pedal by increasing fuel delivery, within the limits allowed by the governor. It is also provided with some means of cutting off fuel delivery when it is wished to stop the engine.

Some kind of governor is associated with the injection pump, either integral with it or attached to it. All vehicle engine governors regulate fuel delivery to control idle speed and maximum speed; the variable-speed governor also regulates intermediate speeds. Operation of the governor may be mechanical or hydraulic, or it may be controlled by manifold depression. Other devices in or attached to the pump include cold start injection advance or fast idle units, turbo boost pressure sensors and anti-stall mechanisms.

These injection pumps are normally very reliable. If they are not damaged by dirt, water or unskilled adjustment they may well outlast the engine to which they are fitted.

Cutaway view of a Bosch PES in-line injection pump

© Robert Bosch Limited

1 Delivery valve holder	9 Control sleeve
2 Spacer	10 Control arm
3 Spring	11 Plunger return
4 Pump barrel	spring
5 Delivery valve	12 Spring seat
6 Inlet/spill port	13 Roller tappet
7 Helix	14 Cam
8 Pump plunger	15 Control rod

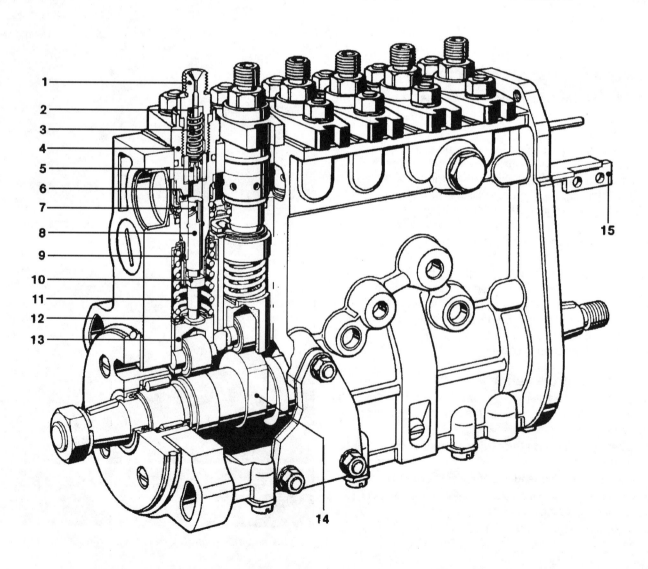

Cutaway view of a Lucas/CAV distribution pump

1 Idle lever
2 Governor main spring
3 Maximum fuel adjuster
4 Speed control (throttle) lever
5 Excess-fuel shaft
6 Metering valve
7 Shut-off valve
8 Hydraulic head
9 Venting orifice
10 Transfer pressure regulating valve
11 Transfer pump
12 Rotor
13 Head-locating fitting and damper
14 Roller and shoe
15 Automatic advance unit
16 Manual advance lever
17 Cam ring
18 Driveshaft rear bearing
19 Governor weight retainer
20 Driveshaft front bearing
21 Driveshaft
22 Governor thrust sleeve
23 Scroll plates
24 Idling spring
25 Fuel return union and residual pressure valve

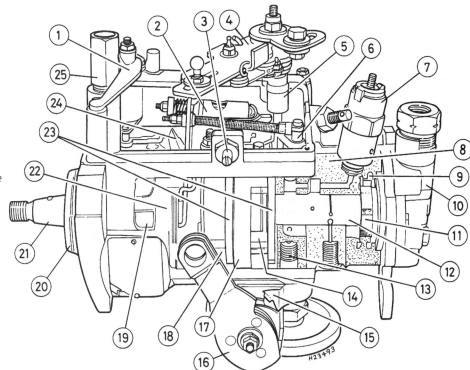

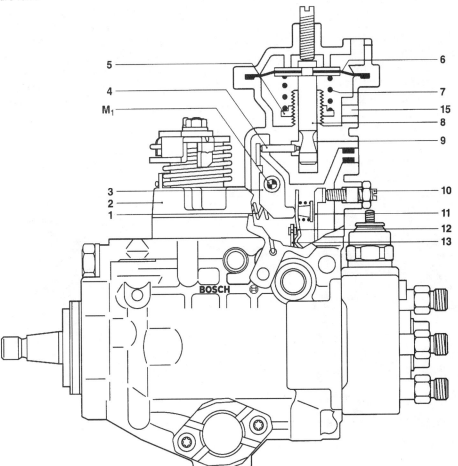

Bosch VE pump with manifold pressure compensation
© Robert Bosch Limited

1 Governor spring
2 Governor cover
3 Stop lever
4 Guide pin
5 Adjusting nut
6 Diaphragm
7 Spring
8 Sliding pin
9 Waisted section
10 Full-load adjusting screw
11 Adjusting lever
12 Tensioning lever
13 Starting lever
M_1 Pivot for stop lever

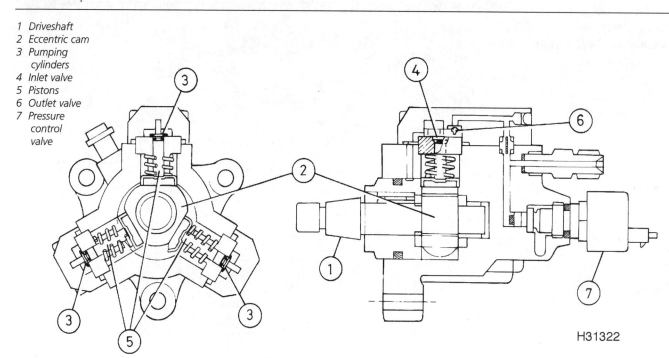

1 Driveshaft
2 Eccentric cam
3 Pumping
   cylinders
4 Inlet valve
5 Pistons
6 Outlet valve
7 Pressure
   control
   valve

H31322

Cutaway view of a high pressure pump - Bosch common rail system

High pressure pumps

On the later common-rail type diesel engines, fuel at very high pressure is supplied to a common reservoir (known as an accumulator rail) by a high pressure pump. The high-pressure pump is most often mounted on the engine in the position normally occupied by a conventional distributor fuel injection pump. The pump is driven at half engine speed by the engine timing belt, timing chain, or possibly by gears, depending on application. The pump is lubricated by the fuel which it pumps.

The high-pressure pump consists of a number of radially-mounted pistons and cylinders *(see illustrations)*. The pistons are operated by an eccentric cam mounted on the pump drive spindle. As a piston moves down, fuel enters the cylinder through an inlet valve. When the piston reaches bottom-dead-centre (BDC), the inlet valve closes, and as the piston moves back up the cylinder, the fuel is compressed. When the pressure in the cylinder reaches the pressure in the accumulator rail, an outlet valve opens, and fuel is forced into the accumulator rail. When the piston reaches top-dead-centre (TDC), the outlet valve closes, due to the pressure drop, and the pumping cycle is repeated. The use of multiple

Common rail high pressure pump fitted (Ford Mondeo shown)

High pressure pump removed (Peugeot 406 2.2 litre)

cylinders (usually three) provides a steady flow of fuel, minimising pulses and pressure fluctuations.

As the pump needs to be able to supply sufficient fuel under full-load conditions, it will supply excess fuel during idle and part-load conditions. This excess fuel is returned from the high-pressure circuit to the low-pressure circuit (to the tank) via the pressure control valve.

The pump incorporates a facility to effectively switch off one of the cylinders to improve efficiency and reduce fuel consumption when maximum pumping capacity is not required (see illustration). When this facility is operated, a solenoid-operated needle holds the inlet valve in the relevant cylinder open during the delivery stroke, preventing the fuel from being compressed.

The engine management ECU receives data from a fuel pressure sensor fitted to the pump, and maintains the required fuel pressure via a pressure control valve integral with the pump.

Replacement guidelines

Note: *For a detailed step-by-step model-specific removal and replacement procedure, refer to the appropriate Service and Repair Manual.*

❑ Because of the very high pressures involved, and the close manufacturing tolerances necessary, absolute cleanliness must be observed whilst working on the fuel system components. After disconnecting fuel pipes from the pump or any other fuel system component, plug or cover the exposed opening to prevent the ingress of dirt (see illustrations).

❑ On common-rail engines, all manufacturers recommend replacing the fuel delivery pipes between the pump and the accumulator rail and the injectors once they have been disturbed, as it is possible for minute metal particles to enter them as a result of tightening the union nuts. If these particles enter the fuel injectors, fuel at high-pressure can enter the combustion chambers unrestricted.

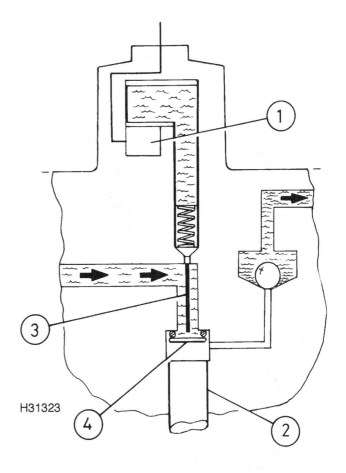

H31323

Cutaway view of a high pressure pump cylinder 'switch-off' mechanism - Bosch common rail system

1 Electromagnet   3 Needle
2 Pumping cylinder   4 Inlet valve

Caps and plugs are available to seal disconnected fuel pipes and components ...

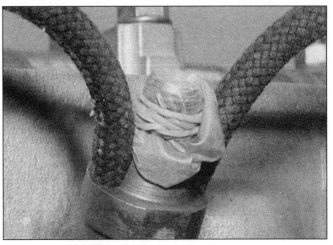

... or use the finger tips from rubber gloves and secure them over the fuel ports with elastic bands

**Don't allow the fuel pump port to rotate when slackening a
fuel union - counterhold it with a second spanner**

Use crow-foot adapters ...

❏ Label the fitted positions of any wiring plugs and hoses
prior to removal.

❏ Allow a few minutes between switching off the engine
and disconnecting any fuel pipes, to allow the residual
fuel pressure to reduce a little.

❏ Do not attempt to remove any component from the high-
pressure pump. According to most manufacturers, should
a component be disturbed, the complete pump must be
renewed - check with your dealer or diesel specialist.

❏ Use a second spanner to counter-hold the union nuts
when slackening a pipe union - do not allow the unions
in the pump to rotate *(see illustration)*.

❏ Tighten the fuel pipe unions to the correct torque using a
crow's-foot spanner *(see illustrations)*.

❏ On common-rail engines, as the injection timing is
controlled by the engine management ECU, checking and
adjusting the pump timing is not necessary.

... and a torque wrench (arrowed) to tighten the unions

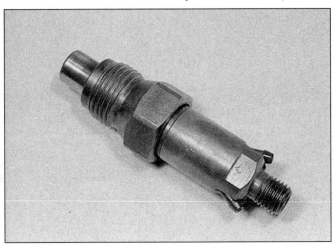

Mechanical injector

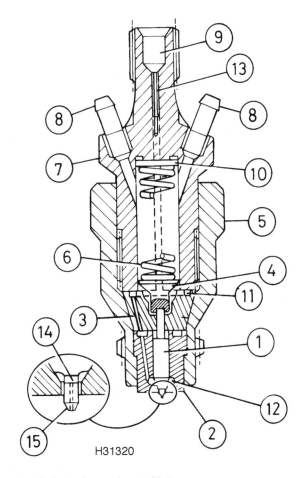

H31320

Sectional view of a mechanical injector

1 Nozzle needle	9 Fuel inlet
2 Nozzle body	10 Pressure-adjusting shim
3 Sleeve	11 Ring groove
4 Spindle	12 Pressure chamber
5 Nozzle retainer	13 Integral filter
6 Spring	14 Transverse bore
7 Body	15 Longitudinal bore
8 Fuel return	

5   Fuel injectors

Description

There are three main designs of injector fitted to diesel engines. On traditional indirect and direct injection engines, the injectors are mechanically operated. When the fuel supplied by the injection pump reaches a certain pressure, the spring loaded valve within the injector is forced from its seat, and fuel is injected into the engine *(see illustrations)*.

On common-rail diesel engines, the injection timing is controlled by the engine management ECU, via electrically operated solenoids incorporated into the injectors *(see illustrations below and overleaf)*. Fuel at the required pressure is supplied to the injectors from the accumulator rail. At the required moment, the ECU operates the solenoid, the spring loaded valve within the injector is forced from its seat and the fuel is injected. In fact, to refine the combustion process and lower exhaust emissions, the system may carry out several pre- or pilot injections, where very small quantities of fuel are injected before and after the main injection. Using these techniques causes a gradual combustion process, lessening the characteristic diesel 'knock', and reduces the emission of harmful exhaust gasses.

On some engines manufactured by the Volkswagen-Audi Group (VAG), a third type of injector is used - the unit or pump injector. On these engines (known as PD for 'Pumpe Duse' engines), fuel is delivered by a camshaft driven lift pump at low pressure to the injectors *(see illustration overleaf)*. No high-pressure or distribution pump is fitted. Instead, a 'roller rocker' assembly, mounted above the camshaft bearing caps, uses an extra set of camshaft lobes to compress the top of each injector once per firing cycle, pressuring the fuel inside the injector. This arrangement creates higher injection pressures (up to 2000 bar). The precise timing of the pre-injection and main injection is controlled by the engine management ECU and a solenoid on each injector. The resultant effect of this system is improved engine torque and power output, greater combustion efficiency, and lower exhaust emissions *(see illustrations overleaf)*.

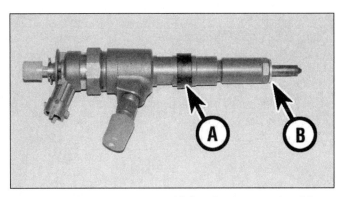

Common rail injector upper seal (A) and copper washer (B)

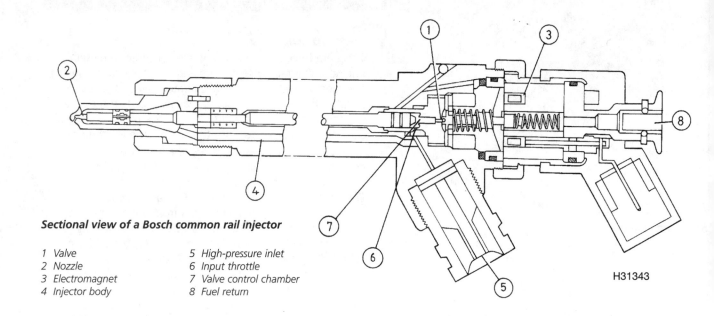

Sectional view of a Bosch common rail injector

1 Valve
2 Nozzle
3 Electromagnet
4 Injector body
5 High-pressure inlet
6 Input throttle
7 Valve control chamber
8 Fuel return

H31343

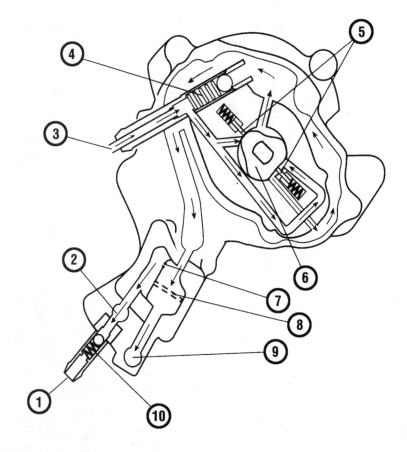

**Cutaway view of a lift pump
fitted to VAG PD engines**

1 Fuel return connection
2 Connection to the fuel return line
  in the cylinder head
3 Connection to the fuel supply line
4 Fuel pressure regulating valve
  (fuel supply)
5 Blocking vanes
6 Rotor
7 Restrictor
8 Strainer
9 Connection to the fuel supply line
  in the cylinder head
10 Pressure regulating valve (fuel
   return)

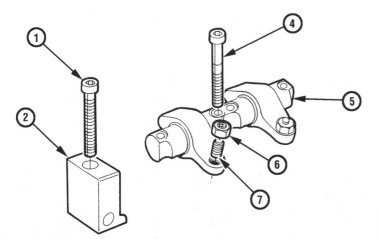

Unit injector fitted to VAG PD engines

1 Bolt
2 Clamping block
3 Cylinder head
4 Bolt
5 Rocker arm
6 Nut
7 Adjuster
8 Unit injector
9 O-ring
10 O-ring
11 O-ring
12 Heat shield
13 Circlip

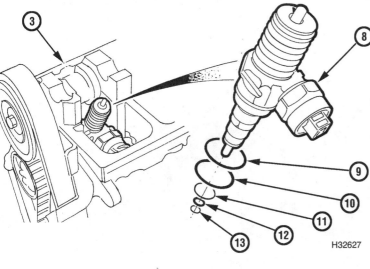

H32627

Cutaway view of a Bosch/VW unit injector

1 Ball-pin
2 Pump piston
3 Piston spring
4 Solenoid valve needle
5 Injector solenoid valve
6 Fuel return line
7 Retraction piston
8 Fuel supply line
9 Injector spring
10 Injector needle damping element
11 Injector needle
12 Heat insulating seal
13 O-rings

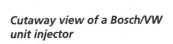

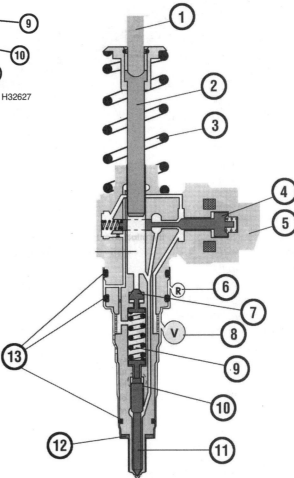

Renew the copper washer ...

... and fire seal on mechanical injectors

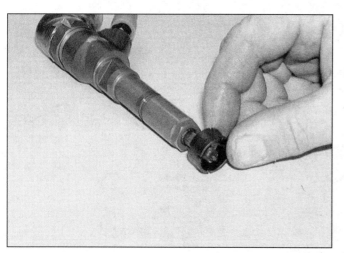

Renew the upper ...

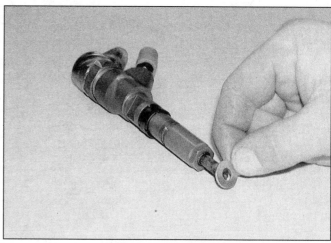

... and lower seal on common rail injectors

Replacement guidelines

Note: *For a detailed step-by-step, model-specific removal and replacement procedure, refer to the appropriate Service and Repair Manual.*

❑ Always renew the seals between the injectors and the cylinder head. Use a little grease to retain the lower sealing washer in place when refitting the injectors *(see illustrations).*

❑ If an injector is stiff to remove, use a pry-bar between the cylinder head and injector - do not apply any force to the solenoid casing.

❑ Do not attempt to clean or dismantle the injectors - have them cleaned and checked by a Diesel injection specialist.

❑ On common-rail engines, the manufacturers insist that the rigid metal pipes between the accumulator rail and the injectors must be replaced once disturbed, as it is possible for minute metal particles to enter them as a

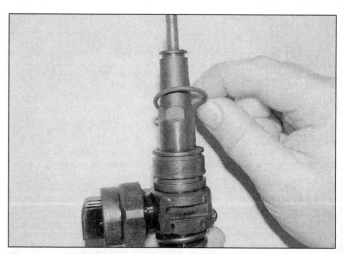

Install the new O-ring seals on unit (pump) injectors without twisting them

result of tightening the union nuts. If these particles enter the fuel injectors, fuel at high-pressure can enter the combustion chambers unrestricted.

❑ On engines with unit injectors, allow the engine to cool completely before starting, as the fuel to and from the injectors runs through galleries in the cylinder head, and therefore gets very hot.

❑ Do not attempt to bend the rigid metal fuel pipes. They are pre-shaped and will be severely weakened if they are misshapened.

❑ After disconnecting fuel pipes from the pump or any other fuel system component, plug or cover the exposed opening to prevent the ingress of dirt.

The crankshaft position sensor (arrowed) may be mounted onto the engine/transmission plate ...

6 Electronic Diesel injection system components

Note: *For a detailed step-by-step, model-specific removal and replacement procedure, refer to the appropriate Service and Repair Manual.*

Crankshaft sensor

Description

Mounted adjacent to the crankshaft, this sensor supplies data regarding the crankshaft speed and position to the engine management ECU. The sensor is basically a magnet with coil wrapped around it. As the signal ring on the crankshaft rotates, the 'teeth' on the ring disturb the sensors magnetic field, creating an electrical signal with a frequency proportional to the engine speed. As one of the teeth on the ring is unique compared to the others (larger, smaller or missing etc.), the ECU can recognise the exact position of the crankshaft. Some systems are sufficiently intelligent to be able to register a cylinder misfire or compression loss, by comparing the rate of crankshaft acceleration/deceleration from one cylinder to the next *(see illustrations)*.

... through the side of the engine block ...

Replacement guidelines

❑ Always tighten the sensor mounting bolts/nuts to the specified torque.

❑ Handle the sensor with care - it is easily damaged.

❑ In order to work reliably, the sensor must be mounted a certain distance away from the signal ring - any dirt or corrosion between the sensor and its mounting location could result in erroneous or missing signals.

... or at the front of the crankshaft

Chapter 3

Airflow meter/sensor

Description

The airflow meter is mounting in the air inlet ducting between the air filter housing and the inlet manifold or turbocharger (as applicable). The meter supplies the engine management ECU with data regarding the amount of air being drawn into the combustion chambers. There are three main designs of meter - moving vane, hot wire, and hot film *(see illustrations)*.

With moving vane meters, the incoming air forces open a hinged vane placed in the air flow - the more air coming in, the further open the vane. Attached to this vane is a potentiometer, which sends a voltage signal proportional to the vane position, to the ECU.

With hot wire meters, the air mass is measured by placing a heated wire in the air stream. The air flowing over the wire would, in theory, cool the wire. However, the control unit (integral with the meter/sensor) ensures the wire temperature does not change by supplying a greater or lesser current to the element as required (a change in temperature affects the wire resistance, which is monitored by the control unit). Therefore the amount of current supplied is directly proportional to the mass of incoming air. Hot film meters work in the same way, but a 'thin film' element is placed in the air stream instead of a wire.

Replacement guidelines

❑ Ensure the seal between the meter housing and the air inlet ducting is in good condition.

❑ Check the condition of the wiring connector pins before condemning a suspect meter.

❑ Do not attempt to clean a hot-wire or hot-film element - they are easily damaged.

Vane type airflow meter ...

... hot wire airflow sensor ...

... and hot film airflow sensor

Fuel injector needle lift sensor

Description

Fitted only to engines with distribution injection pumps and EDC (Electronic Diesel Control), the sensor is integral with the injector, and informs the ECU of the position of the injector needle *(see illustration)*. The ECU uses this information to accurately calculate injection timing and pressure.

Replacement guidelines

❏ The sensor is integral with the injector, and cannot be replaced separately. If faulty, the injector must be replaced.

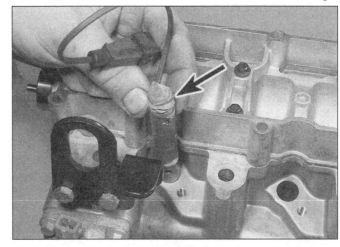

The needle lift sensor (arrowed) is integral with the injector

Coolant temperature sensor

Description

The sensor is fitted to the cylinder head or coolant outlet housing of the engine, where it informs the engine management ECU of the coolant temperature. The resistance of the sensor varies in relation to the coolant temperature, therefore the voltage signal supplied to the ECU increases or decreases as the temperature changes.

There are two types of sensor:; a negative temperature coefficient (NTC) sensor - where the resistance decreases as the temperature increases, and a positive temperature coefficient (PTC) sensor - where the resistance increases as the temperature increases.

Replacement guidelines

❏ Renew the sensor sealing washer/O-ring prior to refitting. Where no sealing washer is fitted, apply a thin layer of RTV sealing compound to the sensor threads *(see illustration)*.

❏ Most sensors are screwed into place, whilst others are retained by a clip *(see illustrations)*.

Always renew the coolant temperature sensor seal (arrowed)

Some sensors are screwed into place ...

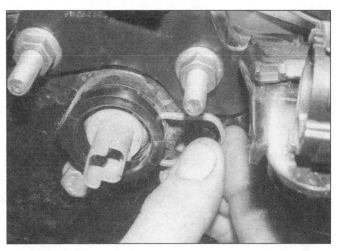

... whilst some are clipped

Fuel temperature sensor

Description

Normally only fitted to common-rail engines, this sensor informs the engine management ECU of the fuel temperature. The ECU can then calculate fuel density, and adjust injection times and pressures accordingly. The data can also be used by the ECU to determine when the returning fuel needs to be diverted through the fuel cooler matrix, prior to entering the (plastic) fuel tank.

As with the coolant temperature sensor, the resistance of the sensor varies according to the fuel temperature.

Replacement guidelines

❏ The sensors are clipped in to the fuel hose manifolds/connecting pieces, or screwed directly into the accumulator rail. Be prepared for fuel spillage as the sensor is removed *(see illustration)*.

❏ On some engines, the sensor is considered an integral part of the accumulator rail, and must not be removed - check in the Service and Repair Manual, or with your dealer/diesel specialist.

Fuel temperature sensor (arrowed)

Fuel quantity servo position sensor

Description

On engines with distributor pumps and EDC (Electronic Diesel Control), this sensor is fitted to the pump, and informs the engine management ECU of the position of the servo which regulates the quantity of fuel injected per firing stroke.

Replacement guidelines

❏ The sensor is integral with the fuel injection pump, and cannot be renewed separately.

Some accelerator pedal position sensors (arrowed) are fitted to the end of the accelerator cable ....

Accelerator pedal position sensor

Description

In order for the engine to respond to the demands made by the driver via the accelerator pedal, common-rail (and some EDC) engines are equipped with a sensor which informs the ECU of the pedal position and rate of change. The ECU can then use this information to adjust the injection timing and pressure to control the engine acceleration or deceleration. This system is often referred to as 'Drive-by-wire'.

The sensor is normally mounted at the top of the accelerator pedal, although on some models, an accelerator cable is still fitted, with the end of the cable attached to a position sensor located in the engine compartment *(see illustrations)*.

Replacement guidelines

❏ On some models, the sensor is integral with the pedal assembly - check prior to removal.

❏ Where an accelerator cable is still fitted (normally Peugeot/Citroen vehicles), carry out the cable adjustment procedure after replacing the sensor.

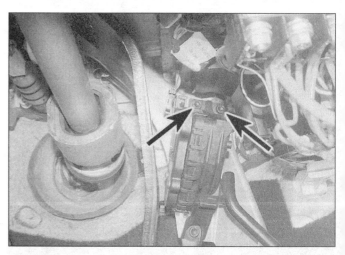

... whilst some are fitted to the accelerator pedal assembly (arrowed)

Manifold absolute pressure (MAP) sensor

Description

This sensor supplies the engine management ECU with data regarding the air pressure/vacuum in the inlet manifold. As the pressure/vacuum is directly proportional to the engine load, the ECU can adjust the injection quantity and timing to suit. Normally, a MAP sensor is only fitted to engines with EDC and a distributor injection pump where no airflow meter is fitted.

Replacement guidelines

❏ The sensor is often fitted to the engine compartment bulkhead, and connected to the inlet manifold by a rubber tube - check for cracks in the tube before condemning a suspect sensor *(see illustration)*.

❏ Where the sensor is fitted directly onto the manifold, renew the rubber seal prior to refitment *(see illustration)*.

The MAP sensor can be connected via a tube ...

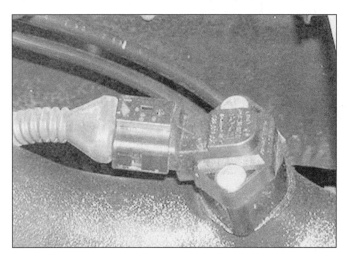

... or fitted directly to the inlet manifold

Intake air temperature sensor

Description

In order to 'fine-tune' the fuel injection timing and quantity, the engine management ECU receives a signal from a sensor located in the air stream in the inlet trunking. When the temperature changes, the sensor resistance changes, and therefore the voltage signal to the ECU. On some later common-rail engines, the ECU also uses this information to determine when the incoming air needs to be heated (by the heater in the air cleaner housing) or cooled (by the intercooler), thus increasing performance and reducing exhaust emissions.

On some engines, the sensor is integral with the air flow meter, and cannot be renewed separately, whilst on others the sensor is fitted to the inlet manifold or trunking *(see illustration)*.

Replacement guidelines

❏ Always renew the sealing washer/O-ring prior to refitting.

❏ Do not attempt to clean the sensor with solvent based cleaners.

Inlet air temperature sensor

Fuel cut-off solenoid

Description

When the ignition key is turned to the 'Off' position, this solenoid is de-energised, and cuts off the fuel supply to the injectors. Fitted to distributor/in-line injection pumps only, the solenoid is often the only means of stopping the engine, and conversely is often the prime suspect should an engine fail to start *(see illustration)*.

Replacement guidelines

❑ Before condemning a suspect solenoid, check for electrical power supply to the solenoid when the ignition switch is turned to the 'On' position.

❑ When operating correctly, the solenoid can be heard to 'click' when the ignition switch is turned on or off.

❑ Renew the sealing washer (where applicable) prior to refitting.

❑ Be careful not to allow dirt into the injection pump whilst replacing the solenoid.

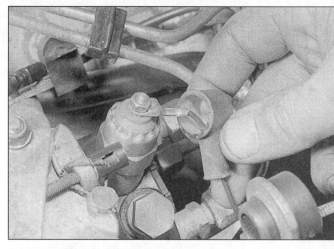

Fuel cut-off solenoid

Turbocharger boost pressure sensor

Description

On some turbocharged engines, this sensor informs the engine management ECU of the air pressure at the turbocharger outlet. As the pressure changes, so does the sensor resistance. The ECU uses this information to control the pressure generated by the turbocharger. It does this by sending signals to the vacuum actuator which determines the turbocharger wastegate position *(see illustrations)*.

Replacement guidelines

❑ Allow the engine to cool completely prior to working on the turbocharger.

❑ Renew the sensor seal prior to refitting.

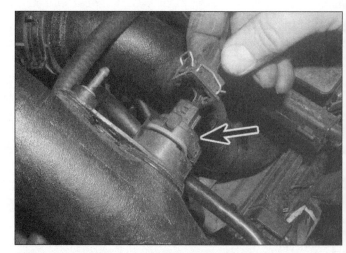

The boost pressure sensor (arrowed) is usually located in the ducting from the turbocharger to the inlet manifold ...

... or bolted directly into the inlet manifold (arrowed)

Engine Management Electronic Control Unit (ECU)

Description

The ECU is the injection system's 'brain'. It receives information from the various sensors, processes this information, then sends signals to various actuators to control the injection system. Stored within the ECU are reference tables known as 'Maps'. As the ECU receives data from the sensors, it compares the values received with those in the Maps, and then 'reads-off' the correct values to send to the actuators (see illustrations).

The ECU knows what values it expects to receive from the various sensors, and should a sensor send a value that is outside the pre-set parameters, or is implausible when compared to data from the other sensors, the ECU will generate a fault code which is stored in its memory. If the fault is sufficiently serious, or emissions related, the ECU will illuminate the MIL (Malfunction Indicator Light) in the instrument cluster, and substitute a known good value for the erroneous signal. When this happens the ECU enters 'Limp-home' or LOS (Limited Operating Strategy) mode, where performance is reduced - but should be sufficient to get you home, or to a workshop. Have the ECU interrogated by a scanner/reader to determine the fault code(s).

Replacement guidelines

❏ Disconnect the battery terminals prior to disconnecting the ECU wiring plug.

❏ Protect the ECU from static electricity. 'Earth' yourself by touching a bare metal section of the engine or body prior to working on the ECU.

❏ Note the fitted locations of any earth/ground wires attached to the ECU mounting bolts/nuts to aid refitment.

❏ Most new ECU's require coding before they will function correctly. This is a process whereby the basic stored values are programmed, and the immobiliser coding is matched to the ECU. This requires specialist equipment. Consult the appropriate Service and Repair Manual or dealer / diesel specialist prior to removing the old ECU.

❏ On some models, once the ECU is fitted, the vehicle needs to be driven for between 10 and 50 miles to enable the basic stored values to be re-learnt. During this period the engine may perform erratically - fast idle speed, occasional stalling etc.

Engine management ECU - Freelander TD4 ...

... Peugeot 206 ...

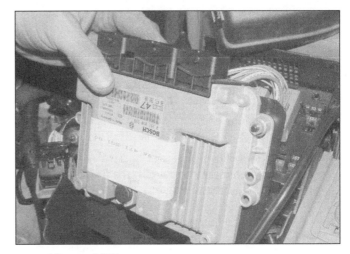

... and Peugeot 307

Inertia fuel cut-off switch

Description

The inertia fuel cut-off switch is designed to cut-off the fuel supply from the tank, should the vehicle be involved in an accident. The sudden deceleration of the impact causes the switch to close. The switch can normally be reset by depressing the button on the top *(see illustration)*.

Replacement guidelines

❏ Be prepared for fuel spillage as the switch is removed.

❏ Cover or plug the fuel pipe openings to prevent dirt ingress.

❏ The switch is often located on the engine compartment bulkhead, or in the passenger cabin - check in the appropriate Service and Repair Manual or Owners Handbook.

To reset the inertia fuel cut-off switch, press the button on the top (arrowed)

Fuel high-pressure sensor

Description

This pressure sensor is normally only fitted to common-rail injection systems. Often screwed directly into the accumulator rail, the resistance of the sensor varies in proportion to the fuel pressure in the rail *(see illustration)*. From the data received from the sensor, the engine management ECU can determine the pressure the pump needs to generate, as well as monitoring its performance.

Replacement guidelines

❏ Most manufacturers insist that the sensor must not be removed from the rail - check prior to attempting removal.

❏ Where the sensor can be removed, always renew the seal.

❏ Be careful not to allow any dirt to enter the accumulator rail - cover or plug the openings.

❏ Tighten the sensor to the specified toque.

The fuel high-pressure sensor (arrowed) is normally fitted to the accumulator rail

Fuel low-pressure sensor

Description

Where fitted (usually to common-rail systems), this sensor informs the engine management ECU of the pressure of the fuel entering the filter assembly. Should the vehicle run low on fuel, and the pressure drop below a pre-determined level, the ECU will stop the engine to prevent damage to the high-pressure pump which relies on the diesel fuel for lubrication.

Replacement guidelines

❏ The sensor is normally integral with the fuel filter housing. If faulty, the complete filter must be replaced.

Fuel pressure control valve

Description

In order to control the pressure generated by the high-pressure pumps fitted to common-rail injection systems, the ECU sends signals to the pressure control valve fitted to the pump *(see illustration)*. If the pressure is excessive, the control valve allows fuel to flow back into the tank. If the pressure is too low, the valve closes, enabling the pump to increase the pressure.

The valve is an electromagnetically-operated ball valve *(see illustration)*. The ball is forced against its seat, against the fuel pressure, by a powerful spring, and also by the force provided by the electromagnet. The force generated by the electromagnet is directly proportional to the current applied to it by the ECU. The desired pressure can therefore be set by varying the current applied to the electromagnet. Any pressure fluctuations are damped by the spring.

Replacement guidelines

❏ Most manufacturers insist that the valve must not be removed from the pump - check prior to attempting removal.

❏ Where the valve can be removed, always renew the seal(s).

❏ Be careful not to allow any dirt to enter the pump - cover or plug the openings.

❏ Tighten the fastenings to the specified toque.

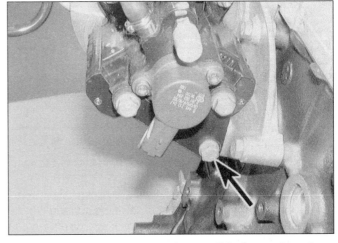

The fuel pressure control valve (arrowed) is mounted on the high-pressure pump

Cutaway view of a pressure control valve - Bosch common rail system

1 Armature
2 Electromagnet
3 Ball valve

4 High pressure inlet
5 Spring

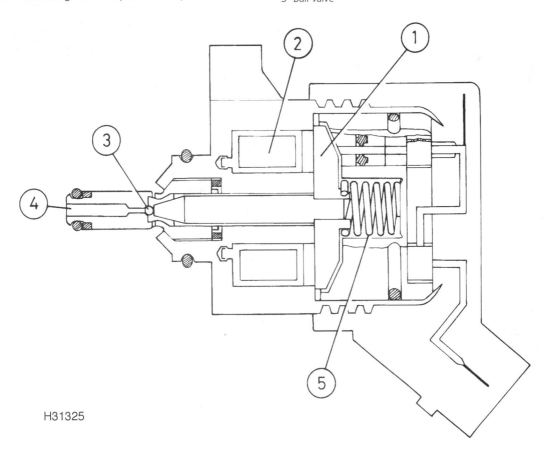

H31325

Chapter 3

Camshaft position sensor

Description

The camshaft position sensor is basically a magnet partially wrapped in a wire coil. The sensor is mounted adjacent to the camshaft, so that as the camshaft rotates, its signal wheel/lug passes over the end of the sensor and disrupts the magnetic field, which creates a frequency signal from the sensor *(see illustration)*. The engine management ECU uses this information to determine the exact firing cycle of the engine, and so determine the injection timing.

Replacement guidelines

❑ As the sensor contains a permanent magnet, take care to remove any metal particles from the sensor casing.

❑ Always renew the seal *(see illustration)*

❑ Tighten the mounting bolt to the specified torque.

❑ With some sensors, it's necessary to set the gap between the sensor tip and the signal wheel - refer to the appropriate Service and Repair Manual *(see illustration)*.

Camshaft position sensor

Away renew the camshaft sensor seal

On some camshaft position sensors, it's necessary to set the gap between the sensor tip and the signal wheel

7 Fuel cooler matrix

Description

Often fitted to common-rail and VAG's PD unit injector engines (see Section 5), the cooler reduces the temperature of the fuel before it returns to the fuel tank. On modern vehicles, the fuel tank is often plastic, and the fuel temperature generated by high-performance diesel engines is sufficient to damage the tank. Additionally, high fuel temperatures can lead to excessive foaming as the fuel enters the tank. A fuel temperature sensor informs the engine management ECU of the fuel temperature, so it can decide to return the fuel straight to the tank, or route it through the cooler matrix (see illustrations).

An alternative type of fuel cooler uses coolant instead of air as the heat exchange medium. This type of cooler is commonly mounted on top of the fuel filter assembly - see Chapter 1.

Fuel cooler matrix - Landrover Freelander TD4 ...

Replacement guidelines

Note: *For a detailed step-by-step, model-specific removal and replacement procedure, refer to the appropriate Service and Repair Manual.*

❏ Be prepared for fuel spillage as the pipes to the cooler are disconnected.

❏ Allow the fuel to cool before disconnecting the pipes.

❏ Plug or cover the pipe/cooler openings to prevent dirt ingress.

... and Peugeot 406

Chapter 3

8   Inlet air heater

Description

On some late common-rail engines, a heater matrix is fitted into the base of the air cleaner housing, to raise the temperature of the incoming air. The matrix is supplied with coolant from the heating system, with the flow being regulated by an electrical valve controlled by the engine management ECU *(see illustration)*.

Heating the incoming air in cold conditions, reduces the amount of time required to bring the engine to operating temperature, which increases performance and reduces the emission of harmful exhaust gases.

Replacement guidelines

Note: *For a detailed step-by-step model-specific removal and replacement procedure, refer to the appropriate Service and Repair Manual.*

❏ Drain the cooling system or clamp the hoses to and from the matrix.

❏ Handle the matrix carefully, the delicate fins are easily damaged.

❏ Renew the seal between the matrix and the air cleaner housing (where applicable)

The air inlet heater is fitted into the base of the air filter housing, and is supplied with engine coolant via a control valve (arrowed)

9   Accumulator rail

Description

The accumulator rail is a reservoir connected between the high-pressure pump and the fuel injectors fitted to common-rail injection systems. Fuel at high pressure is supplied from the pump via rigid metal pipes to the accumulator rail, where the pressure is maintained and supplied to each injector - again, by rigid metal pipes *(see illustrations)*. A fuel pressure sensor and a temperature sensor may be fitted to the rail.

As opposed to in-line/distributor pumps where the injectors each have there own supply from the pump, the accumulator rail is common to all the injectors - hence the term 'common-rail'.

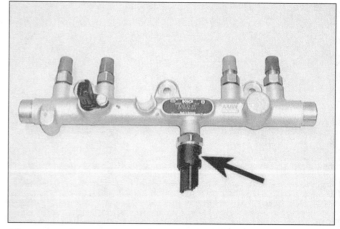

The accumulator rail with the fuel temperature and pressure sensor (arrowed)

Replacement guidelines

Note: *For a detailed step-by-step model-specific removal and replacement procedure, refer to the appropriate Service and Repair Manual.*

❑ Because of the very high pressures involved, and the close manufacturing tolerances necessary, absolute cleanliness must be observed whilst working on the fuel system components. After disconnecting fuel pipes from the accumulator rail or any other fuel system component, plug or cover the exposed opening to prevent the ingress of dirt.

❑ On common-rail engines, all manufactures recommend replacing the fuel delivery pipes between the pump and the accumulator rail and the injectors once they have been disturbed, as it is possible for minute metal particles to enter them as a result of tightening the union nuts. If these particles enter the fuel injectors, fuel at high-pressure can enter the combustion chambers unrestricted.

❑ Tighten the fuel pipe unions to the correct torque using a crow's-foot spanner *(see illustrations)*.

❑ Label the fitted positions of any wiring plugs and hoses prior to removal.

❑ Allow a few minutes between switching off the engine and disconnecting any fuel pipes, to allow the residual fuel pressure to reduce a little.

❑ Do not attempt to remove any component from the accumulator rail without checking with the appropriate Service and Repair Manual or dealer / diesel specialist. According to most manufacturers, should a component be disturbed, the complete pump must be renewed.

❑ Use a second spanner to counter-hold the union nuts when slackening a pipe union - do not allow the unions in the pump, rail or injectors to rotate *(see illustration)*.

Don't allow the fuel pump port to rotate when slackening a fuel union - counterhold it with a second spanner

Use crow-foot adapters ...

Don't allow the fuel pump port to rotate when slackening a fuel union - counterhold it with a second spanner

10   Fuel filter

Description

The fuel filter is an essential part of all diesel injections systems. Due to the high pressures generated by all injection pump, manufacturing tolerances within the pump and injectors are very close. Any dirt in the fuel would very quickly destroy a pump or injector. The filter also prevents water from entering the fuel system by separating it from the fuel, and allowing it to collect at the at the base of the filter housing. As water has very little lubricating qualities, a small quantity can cause a great deal of damage to the pump and injectors.

There are many different types of fuel filter. Some have a replaceable paper element inside a housing bowl; others consist of a disposable canister, either complete with inlet and outlet unions or screwed onto a filter head. Some housings incorporate a heater element to reduce the possibility of the fuel freezing in very cold conditions *(see illustration)*.

Replacement guidelines

Note: *For a detailed step-by-step model-specific removal and replacement procedure, refer to the appropriate Service and Repair Manual.*

❏ Always replace the filter seals. These are normally supplied with the replacement filter *(see illustrations)*.

❏ Drain any accumulated water from the filter housing by opening the drain plug *(see illustration)*.

❏ Be prepared for fuel spillage when disconnecting the fuel inlet and outlet pipes from the filter housing *(see illustration)*.

❏ After fitting a new filter, prime and bleed the fuel system as described in the appropriate Service and Repair Manual.

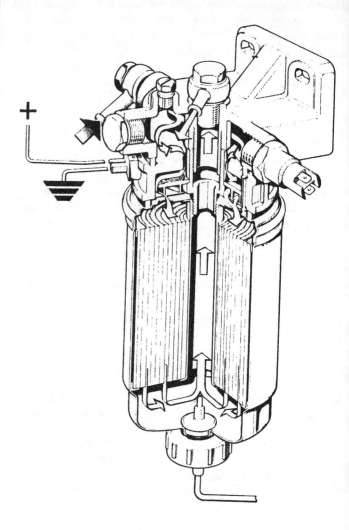

Fuel flow through the filter and heater element

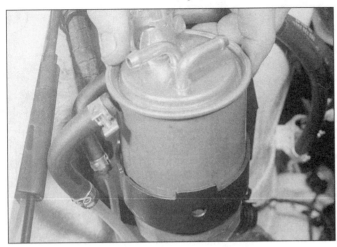

On some vehicles, the complete filter assembly must be replaced ...

... whilst on others the canister unscrews from the housing or 'head'

Always replace the filter seals ...

... and element

Fuel filter drain plug (arrowed)

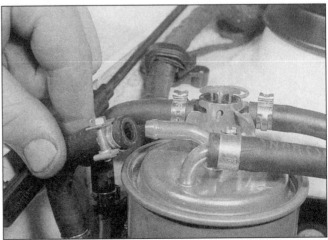

Be prepared for fuel spillage when disconnecting the pipes from the filter/housing

Chapter 3

11 Intercooler

Description

On some turbocharged engines, an intercooler is fitted between the inlet manifold and the turbocharger outlet to cool the incoming air before it enters the engine. Cool air is more dense than hot air, consequently with an intercooler fitted, a higher mass of air is inducted, resulting in greater engine performance.

The intercooler is normally fitted adjacent to the radiator at the front of the engine compartment, where the temperature of the turbocharged air is reduced by the action of the air entering the front of the vehicle as it moves forward (see illustration).

Replacement guidelines

Note: *For a detailed step-by-step model-specific removal and replacement procedure, refer to the appropriate Service and Repair Manual.*

❑ Check the condition of the rubber hoses to and from the intercooler. As the air in these pipes is under pressure any cracking may result in a loss of pressure, and subsequently, engine performance (see illustration).

❑ Take care not to damage the cooling fins of the intercooler, as this will reduce its efficiency.

An intercooler increases the density of the air entering the inlet manifold

Check the condition of the intercooler hoses

12 Fuel distributor pipe

Description

Fitted exclusively to diesel engines with unit- or pump-injectors (see Section 5), the purpose of the distributor pipe is to distribute fuel evenly to the injectors. The pipe is fitted to a passage in the cylinder head, and has a cross-drilling for each pump injector. Fuel supplied by the pump can flow out from the pipe, through the cross-drillings, and into the cylinder head passage which surrounds the pipe. Here, the cool supply fuel mixes with the hot excess (return) fuel forced back into the supply line by the pump injectors *(see illustration)*. This results in the fuel in the distributor pipe being at an even temperature all along the pipe, supplying each pump injector with fuel at the same temperature.

If a distributor pipe was not used, and the fuel was pumped to the pump injectors through a closed pipe, the fuel temperature would rise progressively along the pipe: the hot fuel returned by the pump injectors would be forced towards the end of the pipe by the fuel flowing into the pipe from the lift pump. If the temperature of the fuel varied between pump injectors, differing masses of fuel would be injected by each injector, causing unnecessary engine vibration and stress, and excessively high temperatures in the cylinders receiving 'hot' fuel.

Replacement guidelines

Note: *For a detailed step-by-step model-specific removal and replacement procedure, refer to the appropriate Service and Repair Manual.*

❏ Due to its length, removal of the distribution pipe is only possible once the cylinder head has been removed.

❏ Absolute cleanliness must be observed whilst working on the fuel system components. After disconnecting fuel pipes from the accumulator rail or any other fuel system component, plug or cover the exposed opening to prevent the ingress of dirt.

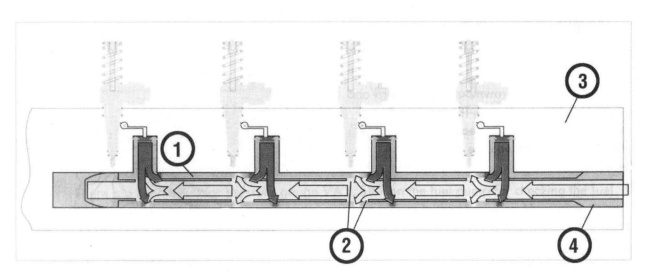

Fuel distributor pipe - VAG PD unit injector engines

1 Annular gap
2 Cross drillings
3 Cylinder head
4 Distributor pipe
5 Fuel from pump injector
6 Fuel to pump injector
7 Mixing of fuel in annular gap

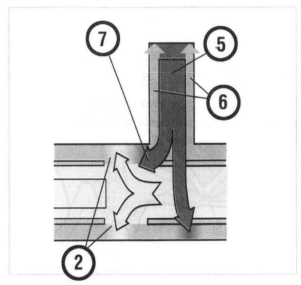

Chapter 3

13 Glow plugs

Description

All indirect injection engines, and most direct injection engines, use glow plugs to assist cold starting. Usually, one plug is used per cylinder (sometimes, one cylinder may not be fitted with a plug due to space restrictions). Some engines still use plugs with an exposed element, but modern engines normally use the sheathed-element plug, where the heating element is enclosed and protected by a metal sheath (see illustrations).

The tips of the glow plugs, which protrude into the pre-combustion or combustion chambers, become red-hot after a few seconds when battery voltage is applied to them. When the injected fuel comes into contact with the hot plug, the fuel ignites.

The time for which the glow plugs need to be energised before the engine can be started will depend on engine temperature and ambient air temperature. The glow plugs are also energised during the starter motor operation, and normally for a few seconds after start-up. The last phase of operation ('afterglow' or 'post heating') contributes to smooth and smoke-free operation of the engine immediately after starting.

The operation of the glow plugs is controlled by the engine management ECU, or a separate pre-heating ECU (see illustrations).

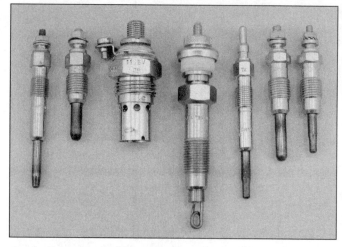

A selection of typical glow plugs

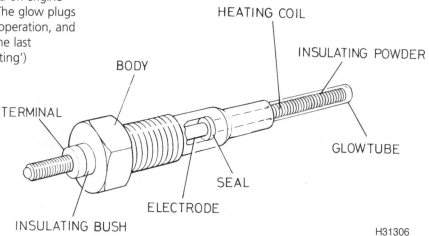

Construction of a sheathed element glow plug

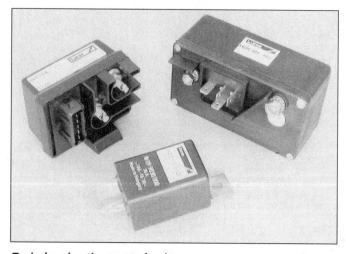

Typical preheating control units

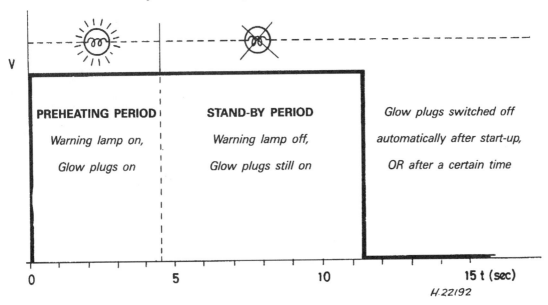

Glow plug voltage (V) versus time for basic automatic preheating system

Testing

Voltage supply checks with a test lamp or voltmeter

Connect the test lamp or voltmeter between the glow plug or flame plug supply wire and earth (vehicle metal) *(see illustration)*. Do not let the live side connections touch earth. Have an assistant energise the preheating system. The test lamp should light brightly, or the voltmeter should read at least 10 volts. (A few Japanese engines use 5-volt glow plugs; on these, a reading of 5 or 6 volts is to be expected.)

If there is no voltage at all, this suggests a fault such as a blown fuse, a disconnected wire, a defective relay, or a defective switch. A blown fuse may only be a pointer to some underlying fault, such as a short-circuit in the wiring or a glow plug which has failed so as to cause a short. The fuse itself may be incorporated in the relay, or it may take the form of a fusible link in the feed wire near the battery.

If the voltage is low and the battery is in good condition, this suggests a bad connection somewhere in the wiring, or possibly a faulty relay.

On systems which control preheating time automatically, a check should now be made of the duration for which the voltage is applied. Remember that on most systems, preheating voltage continues to be applied after the warning lamp has gone out.

Glow plug checks with a test lamp or multimeter

A simple continuity check can be made by disconnecting the wire or metal strap which links the glow plug terminals (ignition off), then connecting the test lamp between the battery positive terminal and each glow plug terminal in turn *(see illustration)*. Alternatively, measure the resistance

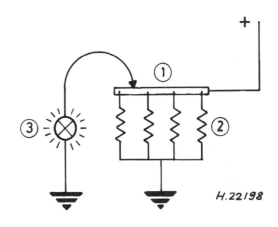

Checking for glow plug supply voltage with a test lamp

1 Glow plug supply wire   3 Test lamp
2 Glow plugs

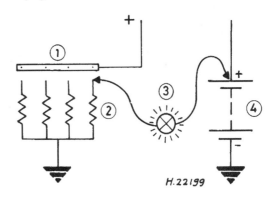

Checking glow plug continuity using a test lamp

1 Glow plug supply wire   3 Test lamp
  (disconnected)   4 Battery
2 Glow plugs

between each glow plug terminal and earth. The lamp should light brightly, or the meter read a very low resistance (typically 1 ohm or less).

If the lamp does not light or the meter shows a high resistance, the glow plug has failed open-circuit, and must be renewed.

The above is only a rough test, and will not detect a glow plug which has failed so as to cause a short-circuit, or one which is no longer heating properly, even though its resistance is still more or less correct. More accurate testing requires the use of an ammeter, of range 0 to 25, or 0 to 30 amps. It should incorporate some kind of overload protection, either in the instrument itself, or by means of a fuse in its lead.

Note: *The procedure which follows applies to glow plugs which operate at full battery voltage, as is the case with most models. It does not apply to the 5-volt glow plugs fitted to some Japanese engines. For testing such plugs, a 6-volt source can safely be used.*

Connect the ammeter between the battery positive terminal and one of the glow plugs. (The glow plugs must still be disconnected from each other.) Note the current draw over a period of 20 seconds or so. Typically, an initial surge of 20 amps or more will fall over 10 to 15 seconds to a steady draw of 9 or 10 amps. A very high draw indicates a short-circuit; a very low draw indicates an open-circuit.

Repeat the current draw check on the other glow plugs, and compare the results. Obvious differences such as a very high or very low draw will not be hard to spot. A difference in the rate at which the current falls off is also significant, and may indicate that the glow plug in question is no longer heating at the tip first.

So far, the tests have concentrated on the electrical condition of the glow plugs. Their physical condition is also important. To establish this, they must be removed and inspected for burning or erosion. Damage can be caused by a fault resulting in too long a post-heating time, but it is more often due to an injector fault *(see illustration)*. If damaged plugs are found, the injectors in the cylinders in question should be removed and checked for spray pattern and calibration.

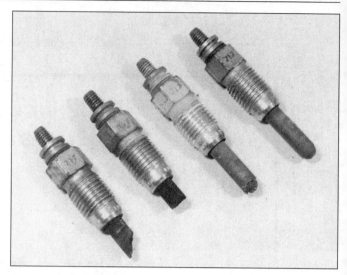

Damaged glow plugs

As a final check, the glow plugs can be energised while they are out of the engine, and inspected to see that they heat evenly. The tip should glow first, with no local hot or cold spots. Some means of supporting the plug while it is being tested must be devised, and the power supply lead should be fused, or should incorporate some other overload protection. Ideally, a purpose-made glow plug tester with a hot test chamber should be used.

Any plug which takes a long time for its tip to glow red, or which shows uneven heating, should be renewed.

Replacement Guidelines

❏ Ensure the glow plugs are cool before attempting removal.

❏ Brush or blow any debris away from the glow plugs prior to removal.

❏ Replace the sealing washers (where fitted).

❏ Handle the glow plugs with care. The heating elements are fragile, and are easily damaged by rough handling.

❏ When refitting, tighten the glow plugs to the specified torque.

Unscrew the nut . . .

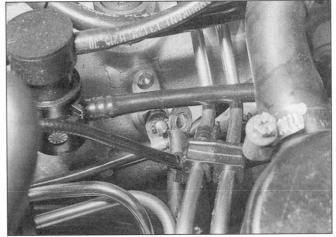

. . . and disconnect the feed wire . . .

Removal, inspection and refitting

Disconnect the battery negative lead.

Remove any surrounding components to allow access to the glow plugs.

Unscrew the nuts from the glow plug terminals, and where applicable recover the washers. Remove the interconnecting wire(s), and where applicable the feed wire from the top of the glow plugs *(see illustrations)*.

Unscrew the glow plugs and remove them from the cylinder head *(see illustration)*.

Inspect the glow plugs for physical damage. Burnt or eroded glow plug tips can be caused by a bad injector spray pattern. Have the injectors checked if this sort of damage is found.

If new plugs are being fitted, ensure that the correct type plug is used, as recommended by the manufacturer.

Refit by reversing the removal operations. Apply a smear of copper-based anti-seize compound to the plug threads, and tighten to the specified torque. Do not overtighten, as this can damage the glow plug element.

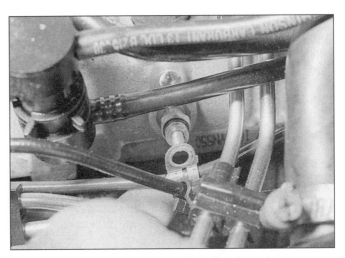

. . . and the interconnecting wire from the glow plug

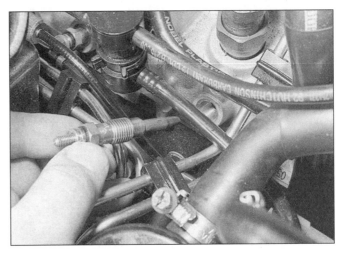

Removing a glow plug

Chapter 3

Alternator mounted vacuum pump (Vauxhall Corsa)

Camshaft driven vacuum pump (Vauxhall Astra)

14 Vacuum pumps

On most diesel engines, the air entering the engine is not throttled. This means there is no vacuum in the inlet manifold, which could be used to power a brake servo. Consequently, a vacuum pump is commonly fitted to supply the servo. The pump maybe driven by the camshaft, or by an auxiliary shaft driven by the timing belt, or may even be integral with the alternator and therefore driven by the auxiliary belt *(see illustrations)*.

The vane pump is the most common type. Here, as the central shaft rotates the vanes sweep the internal chamber(s) evacuating air and passing it through a non-return valve *(see illustration)*. This evacuation results in a vacuum to which the servo is connected. The vacuum source may also be used to supply various other components (EGR valve, Wastegate control actuator, etc.).

In almost all cases, apart from renewing the O-ring seals between the pump and the engine, it is not possible to overhaul the pumps - check in the appropriate Service and Repair manual.

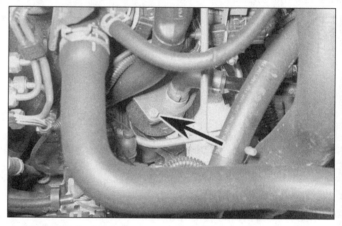

Auxiliary shaft driven vacuum pump (arrowed) (Seat Ibiza)

Removal

Camshaft driven pump

Remove any engine covers as necessary to gain access to the pump, which is normally mounted on the end of the cylinder head, at the opposite end to the timing belt/chain.

1 Outlet
   passage
2 Vane
3 Cells
4 Housing
5 Brake servo
   connection
6 Inlet passage
7 Pump shaft
8 Race

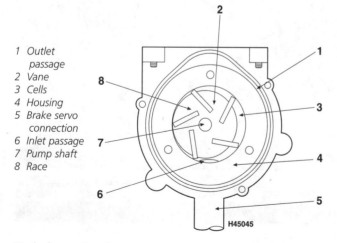

Typical vane-type vacuum pump

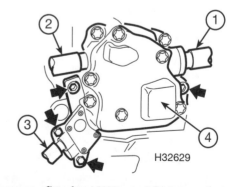

Tandem pump fitted to VW Pump Injector engines

1 Brake servo hose   3 Fuel return hose
2 Fuel supply hose   4 Tandem pump

Disconnect the pipe leading to the servo. Often this pipe is secured by a quick release coupling. If there is more than one pipe connected to the pump, label them to aid refitment. **Note**: On VW Pump Injector engines, the vacuum is generated by a camshaft driven tandem pump, which also supplies fuel at low pressure to the injectors - be prepared for fuel spillage when disconnecting the pipes (*see illustration*).

Undo the bolts/nuts securing the pump to the cylinder head, and remove the pump. Discard the O-ring seals, new ones must be fitted (*see illustration*).

Auxiliary shaft driven pump

Remove any engine covers as necessary to gain access to the pump, which is normally mounted on the front face of the engine block, adjacent to the transmission mounting plate.

Release any retaining clips and disconnect the hose from the top of the pump.

Remove the bolt securing the pump clamp to the engine block, and with draw the pump. Discard the sealing ring, a new one must be fitted (*see illustration*).

Alternator driven pump

This pump is normally fitted to the rear of the alternator. Consequently, access is gained from under the vehicle. Jack up the front of the vehicle and support it securely on axle stands. Remove the engine undershield (where fitted) to gain access to the alternator. Alternatively, it may be preferable to remove the alternator as described in the appropriate Service and Repair Manual.

Disconnect the oil supply, return and vacuum hoses from the pump (*see illustrations*).

Undo the vacuum pump mounting bolts (arrowed) (Peugeot 307 1.4 HDi)

Vacuum pump retaining bolt (arrowed) (Seat Ibiza)

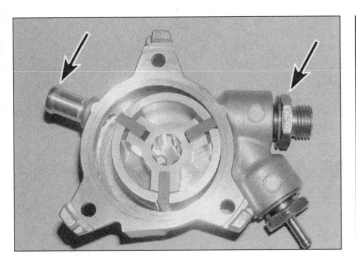

Disconnect the oil supply (A) and return hoses (B) ...

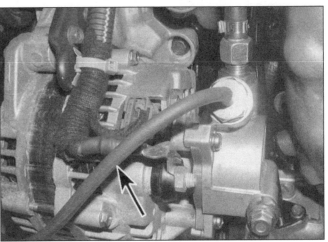

... and the vacuum hose (arrowed)

Undo the bolts and detach the pump from the alternator. Discard the O-ring seal, a new one must be fitted *(see illustration)*.

Overhaul

As no parts appear to be available, overhaul of these units is not possible. If defective, renewal is the only course of action.

Refitting

Refitting these pumps is a reversal of removing them, with the exception of the alternator mounted pump. Prior to refitting this pump, pour 5 cc of clean engine oil into the oil feed aperture.

On all pumps, ensure the mating faces are clean, and always renew the seals. Align the drive lug of the pump with the slot in the end of the camshaft/auxiliary shaft - where applicable *(see illustrations)*.

Check the operation of the braking system before venturing out into traffic.

Undo the bolts and detach the pump from the alternator

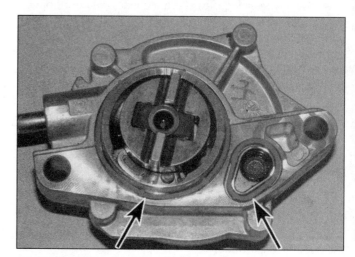

Always renew the pump O-rings (arrowed)

Align the pump drive lug with the slot in the end of the camshaft (arrowed)

Specifications

1 Introduction

The values listed here are not intended as a comprehensive list of Specifications, but are some of the more useful settings needed during routine maintenance, or common repair procedures for popular models with diesel engines. For the full Specifications available, refer to the appropriate Service and Repair Manual.

Chapter 4

	Engine sump drain plug torque (Nm)	Glow plug torque (Nm)	Injector / clamp torque (Nm)	Injector pipe nuts torque (Nm)	Coolant capacity (litres)	Engine oil with filter capacity (Litres)
Alfa Romeo						
155 2.4 JTD	21	28	-	-	6.8	5.0
Audi						
A2 1.2 PD	30	15	12 + 270°	-	5.0	3.8
A2 1.4 PD	30	15	12 + 270°	-	5.0	3.8
A3 1.9 TDi	30	15	20	25	5.0	4.5
A3 1.9 PD	30	15	12 + 270°	-	6.0	4.3
80 1.9 TDi	30	15	20	25	6.5	3.5
A4 1.9 TDi	30	15	20	25	6.5	3.8
A4 1.9 PD	30	15	12 + 270°	-	7.0	3.8
A4 2.5 TDi	25	15	10	30	7.5	5.3
A6 1.9 TDi	30	15	20	25	6.5	3.8
A6 1.9 PD	30	15	12 + 270°	-	7.0	3.8
BMW						
318 TDS	M12: 25 M18: 30 M22: 60	20	65	20	7.5	5.0
320 D	M12: 25 M18: 30 M22: 60	18	10	25	7.0	5.0
325 TDS	M12: 25 M18: 30 M22: 60	20	65	20	9.5	6.7
330 D	M12: 25 M18: 30 M22: 60	18	10	25	-	6.5
520 D	M12: 25 M18: 30 M22: 60	18	10	25	9.4	6.0
525 TDS	M12: 25 M18: 30 M22: 60	20	65	20	9.5	6.7
530 D	M12: 25 M18: 30 M22: 60	18	10	25	-	6.5

	Engine sump drain plug torque (Nm)	Glow plug torque (Nm)	Injector / clamp torque (Nm)	Injector pipe nuts torque (Nm)	Coolant capacity (litres)	Engine oil with filter capacity (Litres)
Chrysler						
Voyager 2.5 CRD	12	-	-	-	14.0	5.3
Cherokee 2.5 D	79	23	68	23	9.5	6.8
Citroën						
AX 1.5 D	10	25	55	20	5.0	4.8
Saxo 1.5 D	10	25	70	20	7.1	4.8
C3 1.4 HDi	16	10	40 + 65°	23	5.6	3.8
ZX1.9 D/TD	30	22	90	20	7.5	4.2
Xsara 1.8D	30	15	90	25	8.8	4.2
Xsara 1.9 DW8	30	22	90	25	8.8	4.2
Xsara 1.9 TD	30	22	90	20	9.0	4.2
Xsara 2.0 HDi	30	10	30	20	8.5	4.3
Xantia 1.9 TD	30	22	90	20	9.0	5.1
Xantia 2.0 HDi	30	10	30	20	8.5	4.3
Xantia 2.1 TD	30	20	90	25	9.0	4.2
C5 2.0 HDi	30	10	30	20	10.7	4.5
C5 2.2 HDi	-	10	40 + 45°	20	10.7	4.7
XM 2.5 TD	30	22	55	20	13.0	8.0
Berlingo 1.9 DW8	30	20	90	25	10.5	4.75
Berlingo 2.0 HDi	30	10	30	20	8.8	4.5
Daewoo						
Musso 2.3 TD	30	20	-	18	10.0	8.0

	Engine sump drain plug torque (Nm)	Glow plug torque (Nm)	Injector / clamp torque (Nm)	Injector pipe nuts torque (Nm)	Coolant capacity (litres)	Engine oil with filter capacity (Litres)
Fiat						
Punto 1.7 TD	-	15	55	30	7.2	4.3
Punto 1.9 JTD	20	15	30	20	6.0	4.3
Brava 1.9 JTD	20	15	30	20	6.3	4.9
Brava 1.9 TD100/75	20	15	55	29	6.5	4.5
Stilo 1.9 JTD	25	-	20	20	5.9	4.7
Marea 1.9 TD100/75	20	15	55	30	6.3	4.5
Marea 1.9 JTD	20	15	20	25	6.0	4.3
Marea 2.4 JTD	20	15	-	25	7.6	5.0
Multipla 1.9 JTD	20	15	30	25	6.3	4.4
Ford						
Fiesta 1.4 TDCi	22	8	30	22	5.5	3.8
Fiesta 1.8 D	25	27	70	25	9.3	5.0
Fiesta 1.8 TD	25	28	70	25	9.3	5.6
Fusion 1.4 TDCi	22	8	30	22	5.5	3.8
Escort 1.8 D	25	25	70	25	8.6	5.0
Escort 1.8 TD	25	25	70	25	9.3	5.1
Focus 1.8 TDCi	25	15	38	35	6.5	5.6
Focus 1.8 TDdi	25	15	23	28	6.5	5.6
Mondeo 1.8 TCi	25	28	70	25	9.3	4.5
Mondeo 2.0 TDCi	25	13	47	40	10.6	6.0
Mondeo 2.0 TDdi	23	13	56	28	10.5	6.5
Galaxy 1.9 TDi	30	15	22	25	9.2	4.3
Galaxy 1.9 PD	30	15	12 + 270°	-	7.2	4.5
Maverick 2.7 TD	-	18	59	23	10.8	7.2
Transit 2.4 Di	23	13	52	28	9.6	7.0
Transit 2.5 TCi	25	-	40	20	11.6	6.0

	Engine sump drain plug torque (Nm)	Glow plug torque (Nm)	Injector / clamp torque (Nm)	Injector pipe nuts torque (Nm)	Coolant capacity (litres)	Engine oil with filter capacity (Litres)
Isuzu						
Trooper 3.0 DT	-	23	7	-	5.8	6.0
Trooper 3.1 DT	83	23	64	29	8.6	6.0
Kia						
Sedona 2.9 CRDi	35	-	20	40	9.4	6.2
Land Rover						
Freelander 2.0 TD4	28	20	10	20	7.2	6.8
Freelander 2.0 TD	25	20	25	20	6.5	4.5
Discovery 200 TDI	45	20	25	20	11.5	6.8
Discovery 300 TDI	35	20	25	20	11.5	6.8
Mercedes						
A-Class A170 CDI	30	15	-	20	6.5	4.5
C-Class C200 CDI	30	12	7 + 90°	20	11.9	6.5
C-Class C270 CDI	30	12	7 + 90°	20	10.6	6.5
E-Class E250 TD	25	20	30	20	9.0	7.0
E-Class E300 TD	25	20	30	20	10.0	8.0
Nissan						
Serena 2.3 D	35	20	65	25	14.5	7.0
Terrano II 2.7 TD	55	18	60	22	10.0	6.2

	Engine sump drain plug torque (Nm)	Glow plug torque (Nm)	Injector / clamp torque (Nm)	Injector pipe nuts torque (Nm)	Coolant capacity (litres)	Engine oil with filter capacity (Litres)
Peugeot						
106 1.5 D	30	22	55	25	6.0	4.5
205 1.7 TD	30	22	90	-	8.3	4.2
206 1.4 HDi	16	9	20	22	-	3.75
206 1.9 D	34	20	90	25	8.2	4.75
206 2.0 HDi	34	10	30	20	6.2	4.5
306 1.9 TD	30	22	90	-	9.0	4.5
306 2.0 HDi	34	10	30	20	6.2	4.5
307 1.4 HDi	16	9	20	22	6.0	3.75
405 1.9 TD	30	25	90	-	7.0	4.5
406 2.0 HDi	30	10	30	22	8.8	4.5
406 2.1 TD	30	25	90	23	9.0	5.2
406 2.2 HDi	-	10	40 + 45°	25	8.8	4.7
Renault						
Clio 1.5 DCi	-	15	28	38	7.3	5.0
Clio 1.9 D	15	20	70	25	7.4	5.2
Clio 1.9 DTi	15	15	27	25	7.5	5.5
Megane 1.9 D	15	20	70	25	7.3	5.5
Megane/Scenic 1.9 DTi	15	15	27	25	7.5	5.5
Megane/Scenic 1.9 DCi	15	15	20	25	7.5	4.6
Laguna 1.9 DTi	15	15	27	-	7.5	5.5
Laguna 1.9 DCi	15	15	20	25	8.2	5.1
Laguna 2.2 DT	-	25	70	25	9.0	7.2
Laguna II 1.9 DTi	15	15	20	30	7.0	4.8
Espace 1.9 DTi	15	15	27	-	8.5	5.1
Espace 2.1 DT	-	25	70	25	8.8	6.5
Espace 2.2 DCi	-	11	-	25	7.5	8.3

	Engine sump drain plug torque (Nm)	Glow plug torque (Nm)	Injector / clamp torque (Nm)	Injector pipe nuts torque (Nm)	Coolant capacity (litres)	Engine oil with filter capacity (Litres)
Rover						
200 2.0 DT	25	20	25	20	7.0	4.8
25 2.0 DT	25	20	25	28	7.0	4.5
MG ZR 2.0 D	25	20	25	28	6.3	4.5
400 2.0 DT	25	20	25	20	7.0	4.0
MG ZS 2.0 D	25	20	25	28	7.0	4.8
600 2.0 DT	25	20	25	20	7.0	5.0
75 2.0 CDT	28	20	10	20	8.2	6.75
800 2.5 DT	55	23	70	18	7.0	6.4
Saab						
9.3 2.2 TiD	18	10	-	25	8.3	5.5
Seat						
Ibiza/Cordoba 1.9 TDi	30	15	20	25	6.0	4.5
Ibiza/Cordoba 1.9 SDi	30	15	20	25	6.0	4.5
Ibiza/Cordoba 1.9 PD	30	15	12 + 270°	-	5.0	4.5
Alhambra 1.9 TDi	30	15	20	25	6.8	4.3
Alhambra 1.9 PD	30	15	12 + 270°	-	6.8	4.5
Skoda						
Fabia 1.9 SDi	30	15	20	25	6.5	4.5
Felica 1.9D	30	25	70	25	6.0	5.0
Octavia 1.9 TDi	30	15	20	25	6.3	4.5
Octavia 1.9 PD	30	15	12 + 270°	-	6.3	4.5
Suzuki						
Vitara/ Grand Vitara 2.0 TD	35	18	65	28	6.6	5.5

	Engine sump drain plug torque (Nm)	Glow plug torque (Nm)	Injector / clamp torque (Nm)	Injector pipe nuts torque (Nm)	Coolant capacity (litres)	Engine oil with filter capacity (Litres)
Toyota						
Corolla 2.0 D	34	13	64	30	7.0	5.1
Corolla 2.0 D-4D	34	13	26	-	7.0	6.4
Carina-E 2.0 DT	34	13	64	30	7.0	5.1
Avensis 2.0 DT	34	13	64	30	6.9	5.1
Avensis 2.0 D-4D	34	13	26	-	7.0	6.4
Previa 2.0 D-4D	34	13	26	-	7.6	5.7
Rav 4 2.0 D-4D	34	13	26	-	6.9	6.4
Vauxhall						
Corsa-C 1.7 Di	78	20	50	25	6.1	4.5
Corsa-C 1.7 DTi	-	18	22	23	7.1	4.5
Astra-F 1.7 DTL	-	20	70	25	7.8	5.0
Astra-F 1.7 D	-	20	70	25	7.8	5.0
Astra-G 1.7 DTi	-	18	22	23	7.1	4.5
Astra-G 1.7 DT	10	20	70	25	8.7	5.5
Astra-G 2.0 DTi	10	10	-	30	7.9	5.5
Cavalier 1.7 DT	78	20	50	25	7.4	4.5
Vectra 1.7 DT	78	20	50	25	6.8	5.0
Vectra 2.0 DTi	10	10	-	25	7.2	5.5
Vectra 2.2 DTi	10	10	-	30	7.2	5.5
Omega 2.2 DTi	10	10	-	30	7.9	5.5
Omega 2.5 DT	M12: 25 M16: 45	22	60	25	10.0	6.5
Omega 2.5 DTi	10	18	9	25	10.2	6.5
Zafira 2.0 DTi	10	10	-	30	7.9	5.5
Frontera 2.2 DTi	10	10	-	30	8.0	5.5
Frontera 2.3 D	-	20	70	25	10.9	5.5
Frontera 2.5 DT	25	15	70	23	8.8	6.2
Frontera 2.8 DT	44	22	37	-	8.8	5.5

	Engine sump drain plug torque (Nm)	Glow plug torque (Nm)	Injector / clamp torque (Nm)	Injector pipe nuts torque (Nm)	Coolant capacity (litres)	Engine oil with filter capacity (Litres)
VW						
Lupo 1.4 PD	30	15	12 + 270°	-	5.0	3.8
Polo 1.7 SDi	30	15	20	25	6.5	4.7
Polo 1.9 SDi	30	15	20	25	6.5	4.3
Polo 1.9 TDi	30	15	20	25	6.5	4.3
Golf 1.9 DT	30	25	70	25	6.5	4.3
Golf/Bora 1.9 TDi	30	15	20	25	6.5	4.3
Golf/Bora 1.9 PD	30	15	12 + 270°	-	6.0	4.5
Passat 1.9 DT	30	25	70	25	7.0	4.3
Passat 1.9 TDi	30	15	220	25	7.5	3.8
Passat 1.9 PD	30	15	12 + 270°	-	7.0	3.8
Passat 2.5 TDi	25	15	10	30	10.0	5.4
Sharan 1.9 TDi	30	15	20	25	6.8	4.3
Sharan 1.9 PD	30	15	12 + 270°	-	6.8	4.5
Volvo						
S40/V40 1.9 DT	Steel: 42 Alum: 20	20	70	23	6.3	5.0
S60 2.4 D5	35	10	10	28	9.0	6.5
850 2.5 TDi	38	15	20	25	12.5	6.0
S/V70 2.5 TDi	38	15	20	25	12.5	6.0
S/V70 2.4 D5	35	10	10	28	9.0	6.5
940/960 2.4 DT	50	22	70	25	11.0	6.7
S80 2.5 TDi	38	15	20	25	12.5	6.0
S80 2.4 D5	35	10	10	28	9.0	6.5

Chapter 4

Fault diagnosis 5

1   Introduction

The majority of starting problems on diesel engines are electrical in origin.

When investigating difficult starting, make sure that the correct starting procedure is understood and is being followed. Some drivers are unaware of the significance of the preheating warning light; many modern engines are sufficiently forgiving for this not to matter in mild weather, but with the onset of winter, problems begin.

As a rule of thumb, if the engine is difficult to start, but runs well when it has finally got going, the problem is electrical (battery, starter motor or preheating system). If poor performance is combined with difficult starting, the problem is likely to be in the fuel system. The low-pressure (supply) side of the fuel system should be checked before suspecting the injectors and injection pump. *Normally the pump is the last item to suspect, since (unless it has been tampered with) there is no reason for it to be at fault.*

Bear in mind that all modern diesel engine vehicles with electronic control have a self-diagnosis system which will store details of any faults as fault codes in the electronic control unit memory (see Section 19). If a fault code is present, a warning light will normally be illuminated on the instrument panel to inform the driver.

Engine won't start

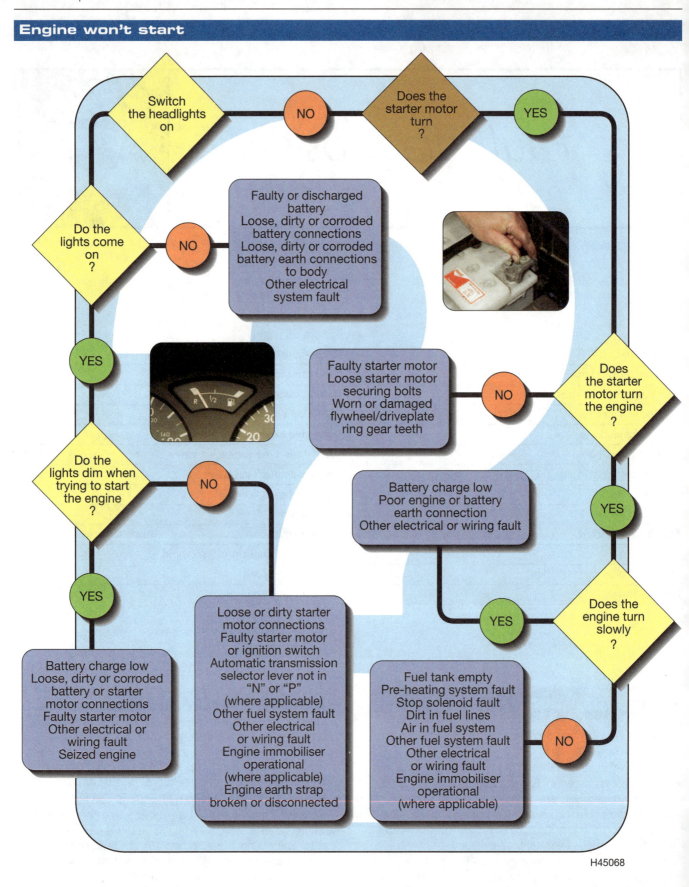

Switch the headlights on

NO → Does the starter motor turn ? **YES**

Do the lights come on ? **NO**

Faulty or discharged battery
Loose, dirty or corroded battery connections
Loose, dirty or corroded battery earth connections to body
Other electrical system fault

YES

Faulty starter motor
Loose starter motor securing bolts
Worn or damaged flywheel/driveplate ring gear teeth

NO → Does the starter motor turn the engine ?

Battery charge low
Poor engine or battery earth connection
Other electrical or wiring fault

YES

Do the lights dim when trying to start the engine ? **NO**

YES → Does the engine turn slowly ?

Battery charge low
Loose, dirty or corroded battery or starter motor connections
Faulty starter motor
Other electrical or wiring fault
Seized engine

Loose or dirty starter motor connections
Faulty starter motor or ignition switch
Automatic transmission selector lever not in "N" or "P" (where applicable)
Other fuel system fault
Other electrical or wiring fault
Engine immobiliser operational (where applicable)
Engine earth strap broken or disconnected

Fuel tank empty
Pre-heating system fault
Stop solenoid fault
Dirt in fuel lines
Air in fuel system
Other fuel system fault
Other electrical or wiring fault
Engine immobiliser operational (where applicable)

NO

H45068

Engine runs poorly

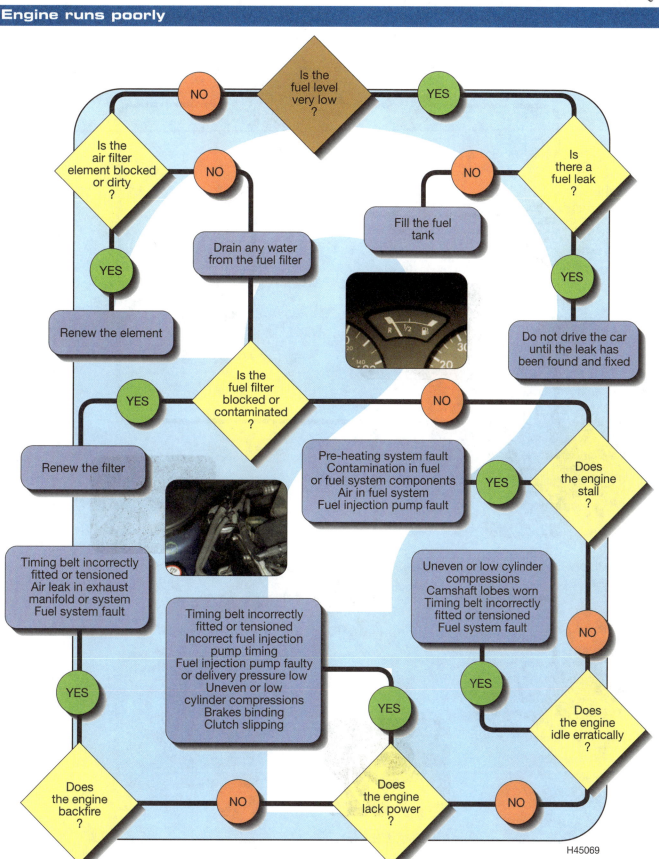

Is the fuel level very low?

NO → Is the air filter element blocked or dirty?

YES → Is there a fuel leak?

NO (from fuel leak) → Fill the fuel tank

YES (air filter) → Renew the element

NO (air filter) → Drain any water from the fuel filter

YES (fuel leak) → Do not drive the car until the leak has been found and fixed

Is the fuel filter blocked or contaminated?

YES → Renew the filter

NO → Does the engine stall?

YES (stall) → Pre-heating system fault
Contamination in fuel or fuel system components
Air in fuel system
Fuel injection pump fault

NO (stall) → Does the engine idle erratically?

YES (filter/renew) → Timing belt incorrectly fitted or tensioned
Air leak in exhaust manifold or system
Fuel system fault

Timing belt incorrectly fitted or tensioned
Incorrect fuel injection pump timing
Fuel injection pump faulty or delivery pressure low
Uneven or low cylinder compressions
Brakes binding
Clutch slipping

Uneven or low cylinder compressions
Camshaft lobes worn
Timing belt incorrectly fitted or tensioned
Fuel system fault

YES (idle erratically) → Uneven or low cylinder compressions...

NO (idle erratically) → Does the engine lack power?

YES → Does the engine backfire?

NO (lack power)

YES (backfire) → Timing belt incorrectly fitted...

NO (backfire) → Does the engine lack power?

H45069

5•3

Engine overheats

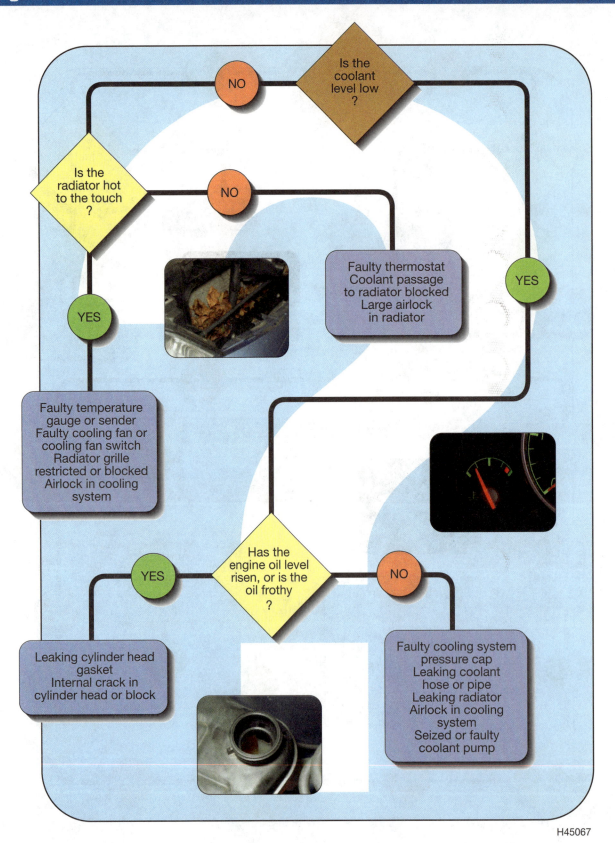

Is the coolant level low ?

NO

Is the radiator hot to the touch ?

NO

YES

YES

Faulty thermostat
Coolant passage to radiator blocked
Large airlock in radiator

Faulty temperature gauge or sender
Faulty cooling fan or cooling fan switch
Radiator grille restricted or blocked
Airlock in cooling system

Has the engine oil level risen, or is the oil frothy ?

YES

NO

Leaking cylinder head gasket
Internal crack in cylinder head or block

Faulty cooling system pressure cap
Leaking coolant hose or pipe
Leaking radiator
Airlock in cooling system
Seized or faulty coolant pump

H45067

Excessive vibration

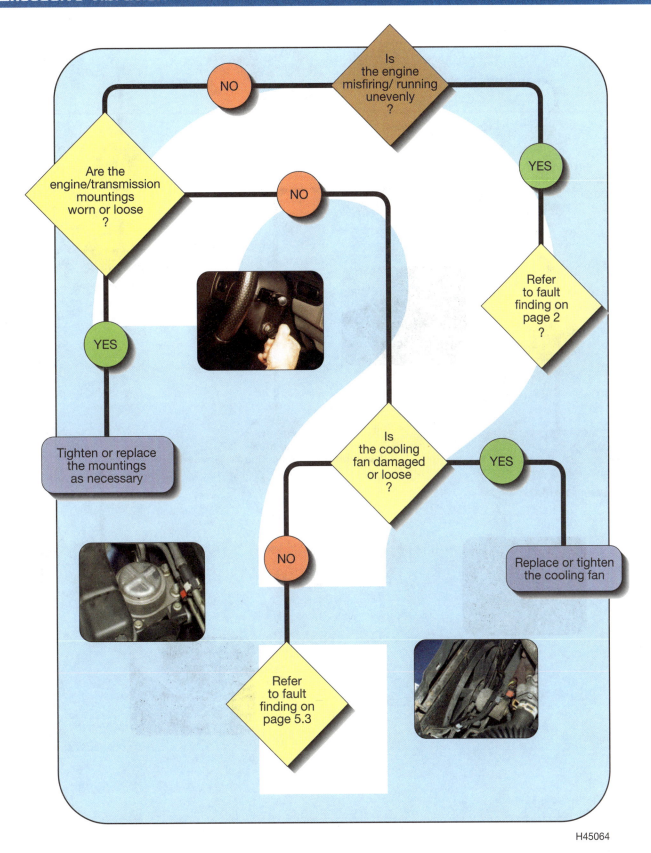

Is the engine misfiring/ running unevenly ?

NO

Are the engine/transmission mountings worn or loose ?

NO

YES

Refer to fault finding on page 2 ?

YES

Tighten or replace the mountings as necessary

Is the cooling fan damaged or loose ?

YES

Replace or tighten the cooling fan

NO

Refer to fault finding on page 5.3

H45064

Chapter 5

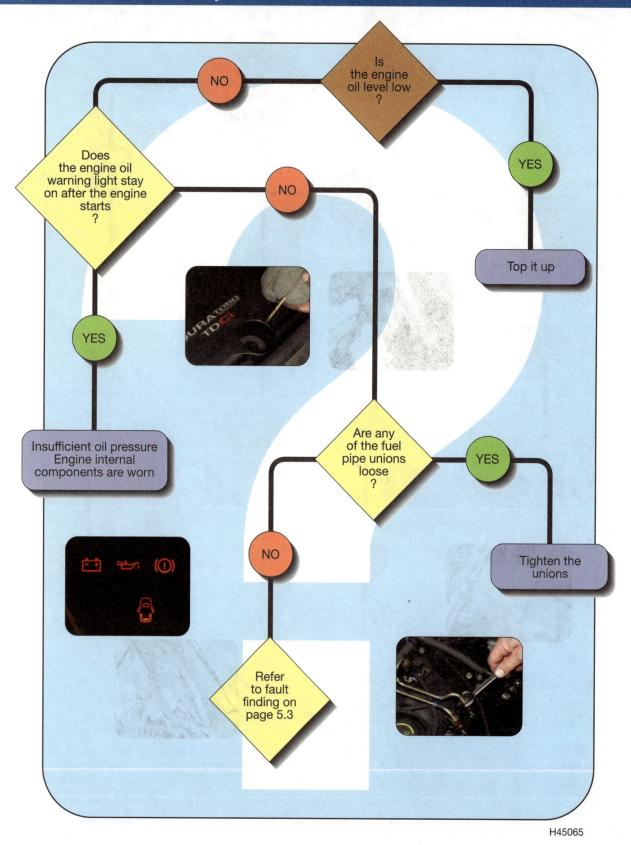

Is the engine oil level low ?

NO

YES

Does the engine oil warning light stay on after the engine starts ?

NO

Top it up

YES

Insufficient oil pressure
Engine internal components are worn

Are any of the fuel pipe unions loose ?

YES

NO

Tighten the unions

Refer to fault finding on page 5.3

H45065

Lack of power

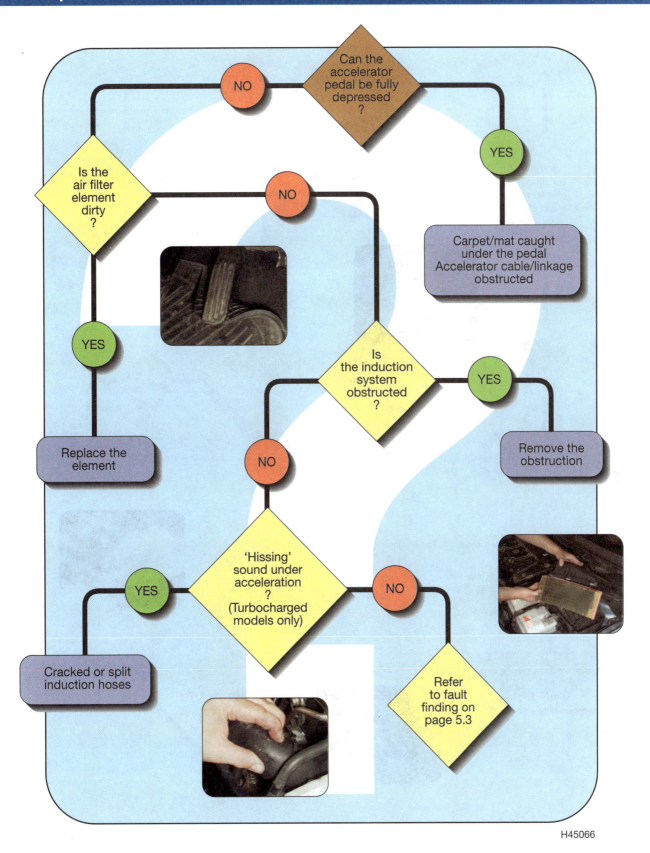

Can the accelerator pedal be fully depressed?

NO

YES → Carpet/mat caught under the pedal Accelerator cable/linkage obstructed

Is the air filter element dirty?

NO

YES → Replace the element

Is the induction system obstructed?

YES → Remove the obstruction

NO

'Hissing' sound under acceleration? (Turbocharged models only)

YES → Cracked or split induction hoses

NO → Refer to fault finding on page 5.3

H45066

Chapter 5

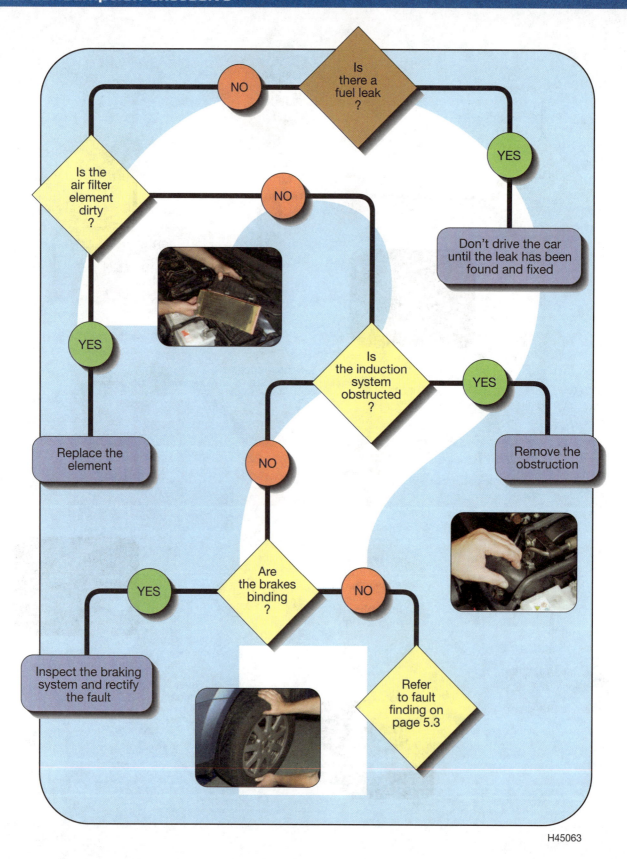

Is there a fuel leak ?

NO

YES

Don't drive the car until the leak has been found and fixed

Is the air filter element dirty ?

NO

YES

Replace the element

Is the induction system obstructed ?

YES

NO

Remove the obstruction

Are the brakes binding ?

YES

NO

Inspect the braking system and rectify the fault

Refer to fault finding on page 5.3

H45063

2   Fault diagnosis checklist

Engine turns but will not start (cold)
☐ Incorrect use of preheating system
☐ Preheating system fault
☐ Fuel waxing (in very cold weather)
☐ Overfuelling or cold start advance mechanism defective
☐ Electronic control system fault

Engine turns but will not start (hot or cold)
☐ Low cranking speed
☐ Poor compression
☐ No fuel in tank
☐ Air in fuel system
☐ Fuel feed restrictions
☐ Fuel contaminated
☐ Engine stop solenoid or mechanism defective
☐ Major mechanical failure
☐ Injection pump internal fault
☐ Electronic control system fault

Low cranking speed
☐ Inadequate battery capacity
☐ Incorrect grade of oil
☐ High resistance in starter motor circuit
☐ Starter motor internal fault

Engine is difficult to start
☐ Incorrect starting procedure
☐ Battery or starter motor fault
☐ Air in fuel system
☐ Fuel feed restriction
☐ Fuel lift pump defective
☐ Poor compression
☐ Valve clearances incorrect
☐ Valves sticking
☐ Blockage in exhaust system
☐ Valve timing incorrect
☐ Injector(s) faulty
☐ Injection pump timing incorrect
☐ Injection pump internal fault
☐ Electronic control system fault

Engine starts but stops again
☐ Fuel very low in tank
☐ Air in fuel system
☐ Idle adjustment incorrect
☐ Fast idle unit fault
☐ Fuel feed restriction
☐ Fuel return restriction
☐ Air cleaner faulty
☐ Blockage in induction system
☐ Blockage in exhaust system
☐ Electronic control system fault
☐ Injector(s) faulty

Engine will not stop when switched off
☐ Stop solenoid defective
☐ Stop actuator leaking or disconnected (vacuum type)
☐ Electronic control system fault

Misfiring/rough idle
☐ Air cleaner dirty
☐ Blockage in induction system
☐ Air in fuel system
☐ Fuel feed restriction
☐ Valve clearances incorrect
☐ Fuel lift pump defective
☐ Valve(s) sticking
☐ Valve spring(s) weak or broken
☐ Poor compression
☐ Overheating
☐ Injector pipe(s) loose, wrongly connected or wrong type
☐ Valve timing incorrect
☐ Injector(s) faulty or wrong type
☐ Injection pump timing incorrect
☐ Injection pump faulty or wrong type
☐ Electronic control system fault

Chapter 5

Lack of power

- ☐ Accelerator linkage not moving through full travel (cable slack or pedal obstructed)
- ☐ Other pump control linkages sticking or maladjusted
- ☐ Air cleaner dirty
- ☐ Blockage in induction system
- ☐ Air in fuel system
- ☐ Fuel feed restriction
- ☐ Fuel lift pump defective
- ☐ Valve timing incorrect
- ☐ Injection pump timing incorrect
- ☐ Blockage in exhaust system
- ☐ Worn camshaft lobes
- ☐ Turbo boost pressure inadequate
- ☐ Valve clearances incorrect
- ☐ Poor compression
- ☐ Injector(s) faulty or wrong type
- ☐ Injection pump faulty or wrong type
- ☐ Electronic control system fault

Fuel consumption excessive

- ☐ External leakage
- ☐ Fuel passing into sump
- ☐ Air cleaner dirty
- ☐ Blockage in induction system
- ☐ Brakes binding
- ☐ Tyre pressures incorrect
- ☐ Valve clearances incorrect
- ☐ Valve(s) sticking
- ☐ Valve spring(s) weak
- ☐ Flame plug leaking fuel (where applicable)
- ☐ Poor compression
- ☐ Valve timing incorrect
- ☐ Injection pump timing incorrect
- ☐ Injector(s) faulty or wrong type
- ☐ Injection pump faulty or wrong type
- ☐ Electronic control system fault

Engine knocks

- ☐ Air in fuel system
- ☐ Fuel grade incorrect or quality poor
- ☐ Injector(s) faulty or wrong type
- ☐ Valve spring(s) weak or broken
- ☐ Valve(s) sticking
- ☐ Valve clearances incorrect
- ☐ Valve timing incorrect
- ☐ Injection pump timing incorrect
- ☐ Piston protrusion excessive/head gasket thickness inadequate (after repair)
- ☐ Valve recess incorrect (after repair)
- ☐ Piston rings broken or worn
- ☐ Pistons and/or bores worn
- ☐ Crankshaft bearings worn or damaged
- ☐ Small-end bearings worn
- ☐ Camshaft worn
- ☐ Timing gears worn
- ☐ Electronic control system fault

Black smoke in exhaust

- ☐ Air cleaner dirty
- ☐ Blockage in induction system
- ☐ Valve clearances incorrect
- ☐ Poor compression
- ☐ Turbo boost pressure inadequate
- ☐ Blockage in exhaust system
- ☐ Valve timing incorrect
- ☐ Flame plug leaking (where applicable)
- ☐ Injector(s) faulty or wrong type
- ☐ Injection pump timing incorrect
- ☐ Injection pump faulty or wrong type
- ☐ Electronic control system fault

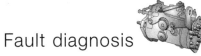

Blue or white smoke in exhaust

- [ ] Engine oil incorrect grade or poor quality
- [ ] Glow plug(s) defective, or control unit faulty (smoke at start-up only)
- [ ] Flame plug leaking (where applicable)
- [ ] Overfuelling device operating after start-up (where applicable)
- [ ] Air cleaner dirty
- [ ] Blockage in induction system
- [ ] Valve timing incorrect
- [ ] Injection pump timing incorrect
- [ ] Injector(s) defective, or heat shields damaged or missing
- [ ] Engine running too cool
- [ ] Oil entering via valve stems
- [ ] Poor compression
- [ ] Head gasket blown
- [ ] Piston rings broken or worn
- [ ] Pistons and/or bores worn
- [ ] Oil consumption excessive
- [ ] External leakage (standing or running)
- [ ] New engine not yet run-in
- [ ] Engine oil incorrect grade or poor quality
- [ ] Oil level too high
- [ ] Crankcase ventilation system obstructed
- [ ] Oil leaking from oil feed pipe into fuel pipe
- [ ] Oil leakage from accessory (vacuum pump, air compressor, etc)
- [ ] Oil cooler leaking into coolant
- [ ] Oil leaking into injection pump (when applicable)
- [ ] Air cleaner dirty
- [ ] Blockage in induction system
- [ ] Cylinder bores glazed
- [ ] Piston rings broken or worn
- [ ] Pistons and/or bores worn
- [ ] Valve stems or guides worn
- [ ] Valve stem oil seals worn
- [ ] Cylinder bores glazed
- [ ] Piston rings broken or worn
- [ ] Pistons and/or bores worn
- [ ] Valve stems or guides worn
- [ ] Valve stem oil seals worn
- [ ] Electronic control system fault

Overheating

- [ ] Coolant leakage
- [ ] Engine oil level too high
- [ ] Electric cooling fan malfunctioning (where applicable)
- [ ] Water pump drivebelt slack or broken
- [ ] Water pump defective
- [ ] Radiator clogged externally
- [ ] Radiator clogged internally
- [ ] Hoses blocked or collapsed
- [ ] Pressure cap defective or incorrect
- [ ] Thermostat defective or incorrect
- [ ] Thermostat missing
- [ ] Air cleaner dirty
- [ ] Blockage in induction system
- [ ] Head gasket blown
- [ ] Cylinder head cracked or warped
- [ ] Valve timing incorrect
- [ ] Injection pump timing incorrect (over-advanced)
- [ ] Injector(s) faulty or wrong type
- [ ] Injection pump faulty or wrong type
- [ ] Electronic control system fault
- [ ] Imminent seizure (piston pick-up)

Crankcase pressure excessive (oil being blown out)

- [ ] Blockage in crankcase ventilation system
- [ ] Leakage in vacuum pump
- [ ] Piston rings broken or sticking
- [ ] Pistons or bores worn
- [ ] Head gasket blown

Erratic running

- [ ] Operating temperature incorrect
- [ ] Stop control or accelerator linkages maladjusted or sticking
- [ ] Air cleaner dirty
- [ ] Blockage in induction system
- [ ] Air in fuel system
- [ ] Injector pipe(s) loose, wrongly connected or wrong type
- [ ] Fuel feed restriction
- [ ] Fuel lift pump defective (where applicable)
- [ ] Valve clearances incorrect
- [ ] Valve(s) sticking
- [ ] Valve spring(s) broken or weak
- [ ] Valve timing incorrect
- [ ] Poor compression
- [ ] Injector(s) faulty or wrong type
- [ ] Injection pump mountings loose
- [ ] Injection pump timing incorrect
- [ ] Injection pump faulty or wrong type
- [ ] Electronic control system fault (where applicable)

Vibration

- [ ] Accelerator linkage sticking
- [ ] Engine mountings loose or worn
- [ ] Cooling fan damaged or loose
- [ ] Crankshaft pulley/damper damaged or loose
- [ ] Injector pipe(s) wrongly connected or wrong type
- [ ] Valve(s) sticking
- [ ] Flywheel or (where applicable) flywheel housing loose
- [ ] Poor (uneven) compression

Low oil pressure

- [ ] Oil level low
- [ ] Oil grade or quality incorrect
- [ ] Oil filter clogged
- [ ] Overheating
- [ ] Oil contaminated
- [ ] Gauge or warning light sender inaccurate
- [ ] Oil pump pick-up strainer clogged
- [ ] Oil pump suction pipe loose or cracked
- [ ] Oil pressure relief valve defective or stuck open
- [ ] Oil pump worn
- [ ] Crankshaft bearings worn

High oil pressure

- [ ] Oil grade or quality incorrect
- [ ] Gauge inaccurate
- [ ] Oil pressure relief valve stuck shut

Injector pipe(s) break or split repeatedly

- [ ] Missing or wrongly-located clamps
- [ ] Wrong type or length of pipe
- [ ] Faulty injector
- [ ] Faulty delivery valve

3   Fuel supply system - testing

It is necessary to test the fuel supply system if it is suspected that air is being drawn into the fuel, or if there is evidence of a blockage causing fuel starvation.

Remember that when a separate fuel lift pump is fitted, the supply lines are under negative pressure between the fuel tank and the lift pump, and under positive pressure from the lift pump to the injection pump. When no separate lift pump is fitted, the supply lines are under negative pressure all the way from the tank to the injection pump. Air can enter at any leaking union, seal, bleed screw or pipe under negative pressure; fuel will not necessarily leak out, even when the engine is stopped.

The fuel return system is also important. On pumps where injection timing is affected by transfer pressure, blockage in the return system can show up as poor performance caused by incorrect timing. The fuel return banjo bolt often incorporates a calibrated orifice; if the supply and return bolts are accidentally interchanged, this too will cause problems (see illustration).

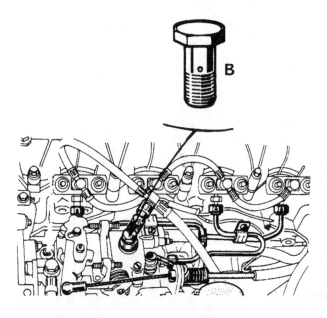

Fuel return banjo bolt (B) with calibrated orifice

Testing for air entry

The presence of air in the fuel can be verified by fitting a piece of transparent hose in the injection pump fuel return line, and running the engine at 2000 to 3000 rpm. If air is being drawn in, bubbles will be visible in the returning fuel. A few bubbles are nothing to worry about, but a continuous stream means trouble. If a return line is fitted at the fuel filter, a similar check can be made there.

Further testing requires a hand-operated vacuum pump, with a vacuum gauge.

Locate the fuel supply line where it leaves the fuel tank. Either disconnect the line and plug it, or (if a flexible hose is fitted at this point) clamp it with a brake hose clamp or self-locking pliers with protected jaws.

Clean around the supply line where it enters the fuel injection pump (or lift pump, if applicable), and disconnect it. Connect the vacuum pump to the line.

 Caution: It is important that no dirt is allowed to enter the pump.

Apply vacuum to the line, and watch the gauge. If the gauge falls, air is entering the line somewhere. If the gauge does not fall, air was being drawn in on the tank side of the point where the line is clamped or plugged. Check that the pick-up pipe is not split.

Disconnect the vacuum pump, and reconnect it at the first union in the direction of the fuel tank (on systems without a lift pump, this will be on the injection pump side of the filter). Again, take care to clean around the union first. Reapply the vacuum and again watch the gauge. If the vacuum is held this time, the leak was in the section first tested.

If the gauge still falls, disconnect the pump and repeat the test one union nearer the fuel tank. Carry on until the leaking union, section or component is located, then repair and re-test. When a diaphragm type hand-priming pump is fitted, do not overlook this as a possible source of air entry, especially if it has seen much service.

Unplug or unclamp the supply line at the tank, and remake the original connections.

Testing for blockage

Systems without a separate lift pump

A fuel line vacuum gauge with the necessary adapters will be needed for this test. The gauge range should be approximately 0 to 1 bar.

Clean around the fuel pipe union on the outlet (injection pump side) of the fuel filter. Connect the gauge to this union using a T-piece.

 Caution: It is important that no dirt is allowed to enter the pump.

Run the engine at maximum rpm, and note the gauge reading. Vacuum of 0.2 bar is acceptable. Any higher reading shows that there is a blockage.

Stop the engine. Remake the original fuel line connections on the pump side of the filter, and connect the vacuum gauge to the inlet (tank side) union of the filter (see illustration). Run the engine at maximum rpm again, and note the gauge reading. Vacuum of up to 0.1 bar is acceptable. Any higher reading shows that there is a blockage on the tank side of the filter.

The difference between the two gauge readings is caused by the resistance to flow of the fuel filter. If the difference is greater than 0.15 bar, this shows that the fuel filter element is clogged, and should be renewed.

A blockage on the tank side of the filter may be caused by one of the following:
   a) Blocked tank vent of filler cap vent (as applicable).
   b) Clogged pick-up strainer (if fitted) in tank.
   c) Tank-to-filter pipe kinked or squashed (external damage).
   d) Tank-to-filter pipe blocked internally.

When testing is complete, stop the engine and remake the original fuel line connections.

Systems with a separate fuel lift pump

The principle of testing is the same as described previously for systems without a lift pump, with the difference being that there is negative pressure on the tank side of the lift

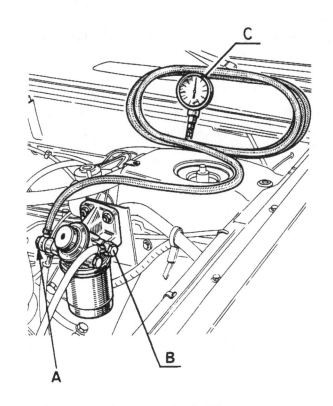

Check the pressure drop across the fuel filter

A Inlet   B Outlet   C Pressure gauge

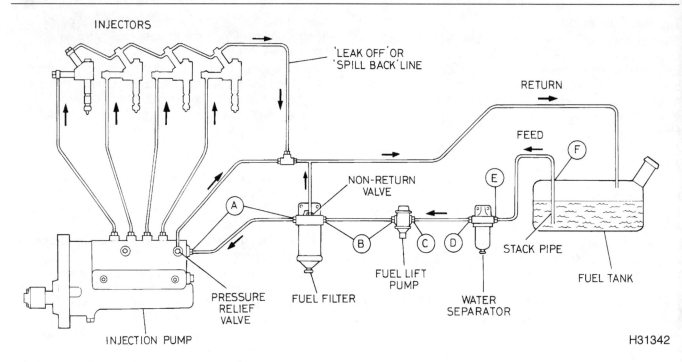

Fuel supply checking points - systems with a separate fuel lift pump

A *Injection pump inlet/filter outlet (lift pump pressure)*
B *Filter inlet/lift pump outlet (lift pump pressure)*
C *Lift pump inlet (vacuum)*

D *Water separator outlet (vacuum)*
E *Water separator inlet (vacuum)*
F *Fuel tank outlet (vacuum)*

pump, and positive pressure on the injection pump side *(see illustration)*. The gauge used must therefore have both positive and negative ranges – typically 0 to 1 bar vacuum, and 0 to 2 bars pressure.

4   Poor compression

Poor compression may give rise to a number of faults, including difficult starting, loss of power, misfiring or uneven running and smoke in the exhaust.

Before looking for mechanical reasons for compression loss, check that the problem is not on the induction side. A dirty air cleaner or some other blockage in the induction system can restrict air intake to the point where compression suffers.

Mechanical reasons for low compression include:
a) *Incorrect valve clearances.*
b) *Sticking valves.*
c) *Weak or broken valve springs.*
d) *Incorrect valve timing.*
e) *Worn or burnt valve heads and seats.*
f) *Worn valve stems and guides.*
g) *Head gasket blown.*
h) *Piston rings broken or sticking.*
i) *Pistons or bores worn.*
j) *Head gasket thickness incorrect (after rebuild).*

Compression loss on one cylinder alone can be due to a defective or badly-seated glow plug, or a leaking injector

sealing washer. Some engines also have a cylinder head plug (for the insertion of a dial test indicator probe when determining TDC), and this should not be overlooked as a possible source of leaks.

Compression loss on two adjacent cylinders is almost certainly due to the head gasket blowing between them. Sometimes the fault will be corrected by renewing the gasket, but a blown gasket can also be an indication that the cylinder head itself is warped. Always check the head mating face for distortion when renewing the gasket. On wet-liner engines, also check the liner protrusion.

Compression test

A compression tester specifically intended for diesel engines must be used, because of the higher pressures involved compared to a petrol engine – see Chapter 6. The tester is connected to an adapter which screws into the glow plug or injector hole. Normally, sealing washers must be used on both sides of the adapter.

Unless specific instructions to the contrary are supplied with the tester, observe the following points:
a) *The battery must be in a good state of charge, the air cleaner element must be clean, and the engine should be at normal operating temperature.*
b) *All the injectors or glow plugs should be removed before starting the test. If removing the injectors, also remove their heat shields (when fitted), otherwise they may be blown out.*

c) *The stop control lever on the injection pump must be operated, or the stop solenoid disconnected, to prevent the engine from running, or fuel from being discharged. Refer to the manufacturer's instructions with regard to testing common rail/pump injector engines.*

There is no need to hold the accelerator pedal down during the test, because the diesel engine air inlet is not throttled. (There are rare exceptions to this case, when a throttle valve is used to produce vacuum for servo or governor operation – for example, early Land Rovers Mercedes.)

The actual compression pressures measured are not so important as the balance between cylinders. Typical values at cranking speeds are:

Good condition	*25 to 30 bar*
Minimum	*18 bar*
Maximum difference between cylinders	*5 bar*

The cause of poor compression is less easy to establish on a diesel engine than on a petrol one. The effect of introducing oil into the cylinders ('wet' testing) is not conclusive, because there is a risk that the oil will sit in the bowl in the piston crown (direct injection engines) or in the swirl chamber (indirect injection engines), instead of passing to the piston rings.

Leakdown test

A leakdown test measures the rate at which compressed air fed into the cylinder is lost. It is an alternative to a compression test, and in many ways is better, since it provides easy identification of where pressure loss is occurring (piston ring, valves or head gasket). However, it does require a source of compressed air.

(!) *Caution: Before beginning the test, remove the cooling system pressure cap.*

This is necessary because if there is a leak into the cooling system, the introduction of compressed air may damage the radiator. Similarly, it is advisable to remove the dipstick or the oil filler cap, to prevent excessive crankcase pressurisation.

Connect the tester to a compressed air line, and adjust the reading to 100% as instructed by the manufacturer.

Remove the glow plugs or injectors, and screw the appropriate adapter into a glow plug or injector hole. Fit the whistle to the adapter, and turn the crankshaft. When the whistle begins to sound, the piston in question is rising on compression. When the whistle stops, TDC has been reached (*see illustrations*).

Engage a gear and apply the handbrake to stop the engine turning. Remove the whistle and connect the tester to the adapter. Note the tester reading, which indicates the rate at which the air escapes. Repeat the test on the other cylinders.

The tester reading is in the form of a percentage, where 100% is perfect. Readings of 80% or better are to be expected from an engine in good condition. The actual reading is less important than the balance between cylinders, which should be within 5% (*see illustration*).

Fitting a leakdown test adapter to a glow plug hole

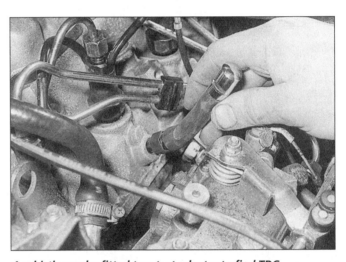

A whistle can be fitted to a test adapter to find TDC

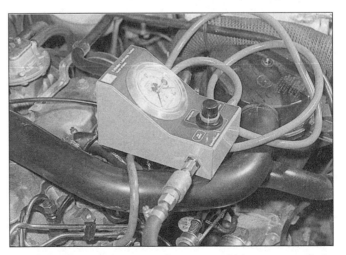

A leakdown tester measures the rate at which compressed air fed into the cylinder is lost

The areas from which escaping air emerges show where a fault lies as follows:

Air escaping from	Probable cause
Oil filler cap or dipstick tube	Worn piston rings or cylinder bores
Exhaust pipe	Worn or burnt exhaust valve
Air cleaner/inlet manifold	Worn or burnt inlet valve
Cooling system	Blown head gasket or cracked cylinder head

Bear in mind that if the head gasket is blown between two adjacent cylinders, air escaping from the cylinder under test may emerge via an open valve in the cylinder adjacent.

5 Air in fuel system

The diesel engine will not run at all, or at best will run erratically, if there is air in the fuel lines. If the fuel tank has been allowed to run dry, or after operations in which the fuel supply lines have been opened, the fuel system must be bled before the engine will run.

Air will also enter the fuel lines through any leaking joint or seal, since the supply side is under negative pressure all the time that the engine is running on models without a fuel tank mounted pump. For testing procedures, see Section 3.

6 Fuel feed restricted

Restriction in the fuel feed from the tank to the pump may be caused by any one of the following faults:
 a) Fuel filter blocked.
 b) Tank vent blocked.
 c) Feed pipe blocked or collapsed.
 d) Fuel waxing (in very cold weather).
Testing of the fuel supply system is covered in Section 3.

Fuel waxing

In the case of fuel waxing, the wax normally builds up first in the filter. If the filter can be warmed, this will often allow the engine to run. Only in exceptionally severe weather will waxing prevent winter-grade fuel from being pumped out of the tank.

 Caution: Do not use a naked flame for this.

Microbiological contamination

Under certain conditions, it is possible for micro-organisms to colonise the fuel tank and supply lines. These micro-organisms produce a black sludge or slime, which can block the filter and cause corrosion on metal parts. The problem normally shows up first as unexpected blockage of the filter.

If such contamination is found, drain the fuel tank, and discard the drained fuel. Flush the tank and fuel lines with clean fuel, and renew the fuel filter; in bad cases, steam-clean the tank as well. If there is evidence that the contamination has passed the fuel filter, have the injection pump cleaned by a specialist.

Further trouble may be avoided by only using fuel from reputable outlets with a high turnover. Proprietary additives are also available to inhibit the growth of micro-organisms in storage tanks or in the vehicle fuel tank.

7 Lack of power

Complaints of lack of power are not always justified. If necessary, perform a road or dynamometer test to verify the condition. Even if power is down, the complaint is not necessarily due to an engine or injection system fault.

Before commencing detailed investigation, check that the accelerator linkage is moving through its full travel. Also make sure that an apparent power loss is not caused by items such as binding brakes, under-inflated tyres, overloading of the vehicle, or some particular feature of operation.

8 Turbo-boost pressure inadequate

If boost pressure is low, power will be down, and too much fuel may be delivered at high engine speeds (depending on the method of pump control). Possible reasons for low boost pressure include:
 a) Air cleaner dirty.
 b) Leaks in induction system.
 c) Blockage in exhaust system.
 d) Turbo control fault (wastegate, actuator, boost sensor).
 e) Turbo mechanical fault.

9 Fuel consumption excessive

Complaints of excessive fuel consumption, as with lack of power, may not mean that a fault exists. If the complaint is justified and there are no obvious fuel leaks, check the same external factors as for lack of power (Section 7) before turning to the engine and injection system.

10 Fuel in the sump

If fuel oil is found to be diluting the engine oil in the sump, this may have arrived in one or more of the following ways:

a) *Down the cylinder bores, especially when the engine is cold.*

b) *Through a leaking fuel lift pump diaphragm (where applicable).*

c) *Through leaking injection pump seals, when these communicate with the timing case.*

d) *Defective injector seal (Unit- or Pump injectors).*

e) *Accumulator rail leaks (where the rail is mounted under the cylinder head cover eg. Isuzu)*

Fuel contamination of the oil can be detected by smell, and in extreme cases, by an obvious reduction in viscosity.

11 Knocking caused by injector fault

On non-common rail or unit/pump injector engines, a faulty injector which is causing knocking noises can be identified as follows:

Clean around the injector fuel pipe unions, then run the engine at a fast idle so that the knock can be heard. Using a suitable spanner, slacken and retighten each injector in turn. (Cover the union with a piece of rag to absorb the fuel which will spray out.)

When the union supplying the defective injector is slackened, the knock will disappear. Stop the engine and remove the injector for inspection (refer to the relevant Service and Repair manual).

On common rail engines if the injector pipes are disturbed, they must be renewed. Have the engine management system ECU self-diagnosis fault memory interrogated before disturbing the pipes.

12 Excessive exhaust smoke

When investigating a complaint of excessive exhaust smoke, check first (by means of a dynamometer or road test) that the smoke is still excessive when the engine has reached normal operating temperature. A cold engine may produce some blue or white smoke until it has warmed up; this is not necessarily a fault.

Black smoke is produced by incomplete combustion of the fuel in such a way that carbon particles (soot) are formed. Incomplete combustion shows that there is a lack of oxygen, either because too much fuel is being delivered, or because not enough air is being drawn into the cylinders. A dirty air cleaner is an obvious cause of air starvation; incorrect valve clearances (where applicable) or worn camshaft lobes should also be considered. Combustion may also be incomplete because the injection timing is incorrect (too far retarded), or because the injector spray pattern is poor.

Blue smoke is produced either by incomplete combustion of the fuel, or by burning engine (sump) oil. This type of incomplete combustion may be caused by incorrect injection timing (too far advanced), by defective injectors, or by damaged or missing injector heat shields.

All engines burn a certain amount of engine oil, especially when cold, but if enough is being burnt to cause excessive exhaust smoke, this suggests that there is a significant degree of wear or some other problem.

White smoke (not to be confused with steam) is produced by unburnt or partially-burnt fuel appearing in the exhaust gases. Some white smoke is normal during and immediately after start-up, especially in cold conditions. Excessive amounts of white smoke can be caused by a preheating system fault, by incorrect injection pump timing, or by too much fuel being delivered by the injection pump (overfuelling device malfunctioning). The use of poor quality fuel with a low cetane number, and thus a long ignition delay, can also increase emissions of white smoke.

Accurate measurement of exhaust smoke requires the use of some kind of smoke meter; these are described in Chapter 6.

13 Oil entering the engine via valve stems

Excessive oil consumption due to oil passing down the valve stems can have three causes:

a) *Valve stem wear.*

b) *Valve guide wear.*

c) *Valve stem oil seal wear.*

In the first two cases, the cylinder head must be removed and dismantled so that the valves and guides can be inspected and measured for wear.

In the case of worn valve stem oil seals, on some engines, these can be renewed without removing the head - refer to the relevant Service and Repair manual. Whether or not this is worthwhile will depend on how worn the valve stems are.

14 Oil consumption excessive

When investigating complaints of excessive oil consumption, make sure that the correct level-checking procedure is being followed. If insufficient time is allowed for the oil to drain down after stopping the engine, or if the level is checked while the vehicle is standing on a slope, a false low reading may result. The unnecessary topping-up which follows may itself cause increased oil consumption, as a result of the level being too high.

Chapter 5

15 Cylinder bore glazing

Engines which spend long periods idling can suffer from glazing of the cylinder bores, leading to high oil consumption, even though no significant wear has taken place. The same effect can be produced by incorrect running-in procedures, or by the use of the incorrect grade of oil during running-in. The remedy is to remove the pistons, deglaze the bores with a hone or 'glaze buster' tool, and to fit new piston rings.

16 Overheating

Complaints of overheating should first be verified, if they are based only on gauge readings, and not on more definite symptoms. Road-test the vehicle, and use a thermometer of known accuracy to measure the temperature of the coolant in the radiator or expansion tank when the gauge shows that overheating is taking place.

 Warning: Take care to avoid scalding when removing the coolant filler cap on a hot engine.

Sometimes the thermostat is removed if it is suspected of being the cause of overheating. If the thermostat is of the bypass-blanking type, this will actually make matters worse, since removing the thermostat increases coolant flow through the bypass, at the expense of flow through the radiator. **Do not** run an engine without the thermostat fitted if it is of this type.

17 Oil contamination

Oil contamination falls into three categories: dirt, sludge and dilution.

Dirt or soot builds up in the oil in normal operation, and is not a problem if regular oil and filter changes are carried out. If it gets to the stage where it is causing low oil pressure, change the oil and filter immediately.

Sludge occurs when inferior grades of oil are used, or when regular oil changing has been neglected; it is more likely to occur on engines which rarely reach optimum operating temperature. If sludge is found when draining, a flushing oil may be used if the engine manufacturer allows it.

The engine should then be refilled with fresh oil of the correct grade, and a new oil filter be fitted.

 Caution: Some engine manufacturers – for example Renault – forbid the use of flushing oil, because it cannot all be drained afterwards.

Dilution is of two kinds: fuel or water (coolant). In either case, if the dilution is bad enough, the engine oil level will appear to rise with use. The routes by which fuel may get into the sump are explained in Section 10.

Coolant dilution of the oil is indicated by the 'mayonnaise' appearance of the oil-and-water mixture. Sometimes oil will also appear in the coolant. Possible reasons are:
  a) *Blown head gasket.*
  b) *Cracked or porous cylinder head or block.*
  c) *Cylinder liner seal failure (on wet-liner engines).*
  d) *Leaking oil-to-coolant oil cooler (when fitted).*
With either type of dilution, the cause must be dealt with, and the oil and filter changed.

18 Engine stop (fuel cut-off) solenoid – emergency repair

Most small diesel engines (not common-rail or pump injector) have a solenoid valve for cutting off the supply of fuel to the high-pressure side of the injection pump when the 'ignition' is switched off. If the solenoid fails electrically or mechanically so that its plunger is in the 'shut' position, the engine will not run. (One possible reason for such a failure is that the 'ignition' has been switched off while the engine speed is still high. In such a case, the plunger will be sucked onto its seat with considerable force, and may jam.)

Should the valve fail on the road and a spare not be immediately available, the following procedure will serve to get the engine running again. (Note however that on some later models the solenoid housing is armoured as an anti-theft measure, and professional assistance will certainly be necessary to get at it.)

 Caution: It is important that no dirt is allowed to enter the injection pump, via the solenoid hole.

With the 'ignition' off, disconnect the wire from the solenoid, and thoroughly clean around the solenoid where it screws into the pump *(see illustration).*

The stop solenoid wiring is secured by a nut (arrowed)

Unscrew the solenoid and remove it. If a hand-priming pump is fitted, operate the pump a few times while lifting out the solenoid, to wash away any particles of dirt. Do not lose the sealing washer.

Remove the plunger from the solenoid (or from the recess in the pump, if it is stuck inside). Refit the solenoid body, making sure the sealing washer is in place, again operating the priming pump at the same time to flush away dirt *(see illustrations)*.

Tape up the end of the solenoid wire so that it cannot touch bare metal.

The engine will now start and run as usual, but it will not stop when the 'ignition' is switched off: it will be necessary to use the manual stop lever (if fitted) on the injection pump, or to stall the engine in gear.

Fit a new solenoid and sealing washer at the earliest opportunity.

19 Reading fault codes

All modern diesel engines with electronic diesel engine control have a self-diagnostic system. If a fault occurs in any of the system sensors or actuators, this is recognised by the electronic control unit (ECU), which stores an appropriate fault code in its memory.

Remove the stop solenoid plunger from the pump

If a sensor or actuator is faulty, the ECU will usually substitute an 'emergency' value in place of the signal normally associated with the relevant sensor or actuator. This will enable the engine to carry on operating, albeit with reduced performance and efficiency. This situation is often referred to as 'limp home' mode.

Stop solenoid components

The 16-pin diagnosis connector may have a cover to unclip ...

... or be located behind a facia trim panel

If a fault code (or codes) is recorded, the ECU will usually illuminate an engine warning light on the instrument panel to inform the driver that there is a problem. The warning light will normally extinguish once the fault code has been read.

To read fault codes, a suitable fault code reader will be required. A diagnostic connector is provided in the vehicle wiring loom, and a fault code reader can be connected via an appropriate adapter harness. In the last few years, the diagnostic connector configuration and location have become somewhat standardised. The 16-pin connector must

be located within arm's reach of the driver's seat *(see illustrations)*. On many vehicles it's necessary to remove a piece of trim or ashtray to access the plug.

Modern common-rail and pump injector engine ECUs also provide 'live data' facility, where the actual outputs from the sensors, and actuator signals from the ECU, can be displayed on the screen of the fault code scanner. This information, combined with any fault codes stored, can be of great assistance in the diagnosis of modern diesel engine running faults.

Tools and equipment

Chapter 6

1 Normal workshop tools

The decision as to what range of tools is necessary will depend on the work to be done, the range of vehicles which it is expected to encounter, and (not least) the financial resources available. The tools in the following list, with additions as necessary from the various categories of diesel-specific tools described later, should be sufficient for carrying out most routine maintenance and repair operations.

> Combination spanners (see text)
> Socket spanners (see text)
> Ratchet, extension piece and universal joint (for use with sockets)
> Torque wrench
> Angle-tightening indicator (see text)
> Adjustable spanner
> Set of sump drain plug keys
> Strap or chain wrench (for fuel and oil filters)
> Oil drain tray
> Feeler gauges
> Pliers
> Long-nosed pliers
> Self-locking pliers ('Mole' wrench)
> Screwdrivers (large and small, flat blade and cross-blade)
> Set of Allen keys
> Set of splined and 'Torx' keys and sockets (see text)
> Ball pein hammer
> Soft-faced hammer
> Puller (universal type with interchangeable jaws)
> Cold chisel
> Scriber
> Scraper
> Centre-punch
> Hacksaw
> File
> Steel straight-edge
> Axle stands and/or ramps
> Trolley jack
> Inspection light
> Inspection mirror
> Telescopic magnet/pick-up tool

Socket and spanner sizes

A good range of open-ended and ring spanners will be required. Most modern vehicles use metric size fastenings throughout, but some early UK-built vehicles may have Imperial fastenings – or a mixture of both.

Split ring spanners (also known as flare nut spanners) are particularly useful for dealing with fuel pipe unions, on which a conventional ring spanner or socket cannot be used because the pipe is in the way. The most common fuel union sizes are 17 mm and 19 mm.

Sockets are available in various drive sizes. The half-inch square drive size is most widely used, and can be used with most torque wrenches. The 3/8 in square drive is also useful for lower torque applications, especially in confined spaces, and 1/4 in and 3/4 in drive tools are also available.

Box spanners should not be overlooked. Box spanners are cheap, and will sometimes serve as a substitute for a deep socket, though they cannot be used with a torque wrench, and are easily deformed.

Angle-tightening

For fastenings such as cylinder head bolts, many manufacturers specify tightening in terms of angular rotation rather than an absolute torque. After an initial 'pre-tightening' torque wrench setting, subsequent tightening stages are specified as angles through which each bolt must be turned. Variations in tightening torque which could be caused by the presence or absence of dirt, oil, etc, on the bolt threads are eliminated. A further benefit is that there is no need for a high-range torque wrench.

The owner/mechanic who expects to use this method of tightening only once or twice in the life of the vehicle may be content to make up a cardboard template or mark the bolt heads with paint spots, to indicate the angle required. Greater speed and accuracy will result from using one of the many angle-tightening indicators commercially available (see illustration).

Splined and 'Torx' bolt heads

The conventional hexagon head bolt is being replaced in many areas by the 'splined' or 'Torx' head bolt. This type of bolt has multiple splines in place of the hexagon: splined bolts generally have 12 splines, and Torx bolts have six splines. A set of splined or Torx keys will be needed to deal with female fixings. Torx bolts with male heads also exist, and for these Torx sockets will be needed.

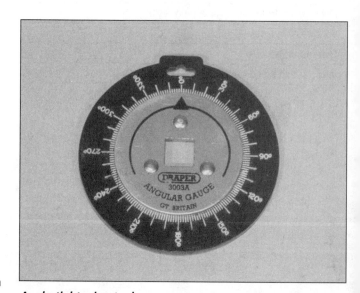

Angle-tightening tool

2 Diesel-specific tools

Basic tune-up and service

Besides the normal range of spanners, screwdrivers and so on, the following tools and equipment will be needed for basic tune-up and service operations on most models:

Deep socket for removing and tightening screw-in injectors

Optical or pulse-sensitive tachometer

Electrical multimeter, or dedicated glow plug tester

Compression or leakdown tester

Vacuum pump and/or gauge

Crow-foot adaptors for slackening and tightening injector/accumulator pipes

Injector socket

The size most commonly required is 27 mm AF; some Japanese injectors require 22 mm AF. The socket needs to be deep in order not to foul the injector body, and on some engines it also needs to be thin-walled *(see illustration).*

Tachometer

The type of tachometer which senses ignition system HT pulses via an inductive pick-up cannot be used on diesel engines, unless a suitable timing light adapter is available. If the engine is fitted with a TDC sensor and a diagnostic socket, an electronic tachometer which reads the signals from the TDC sensor can be used.

Not all engines have TDC sensors; on those which do not, the use of an optical or pulse-sensitive tachometer is necessary.

The optical tachometer registers the passage of a paint mark or (more usually) a strip of reflective foil placed on the crankshaft pulley. It is not so convenient to use as the electronic or pulse-sensitive types, since it has to be held so that it can 'see' the pulley, but it has the advantage that it can be used on any engine, petrol or diesel, with or

Injector socket

without a diagnostic socket *(see illustration).*

The pulse-sensitive tachometer uses a transducer similar to that needed for a timing light. The transducer converts hydraulic or mechanical impulses in an injector pipe into electrical signals, which are displayed on the tachometer as engine speed.

Some dynamic

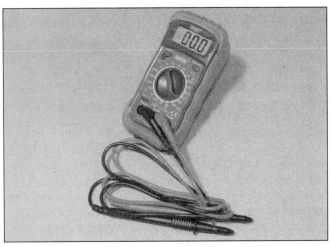

Typical optical tachometer

timing equipment for diesel engines incorporates a means of displaying engine speed. If this equipment is available, a separate tachometer will not be required.

Electrical multimeter or glow plug tester

It is possible to test glow plugs and their control circuitry with a multimeter, or even (to a limited extent) with a 12-volt test lamp. A purpose-made glow plug tester will do the job faster, and is much easier to use *(see illustrations).*

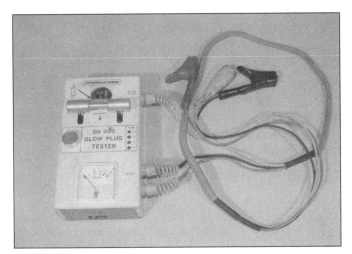

Typical glow plug tester

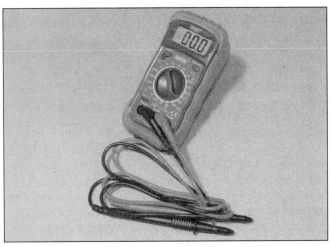

Multimeter

If it is decided to purchase a multimeter, make sure that it has a high current range – ideally 0 to 100 amps – for checking glow plug current draw. Some meters require an external shunt to be fitted for this. An inductive clamp connection is preferred for high current measurement, since it can be used without breaking into the circuit. Other ranges required are dc voltage (0 to 20 or 30 volts is suitable for most applications) and resistance. Some meters have a continuity buzzer in addition to a resistance scale.

Compression tester

A tester specifically intended for diesel engines must be used *(see illustration)*. The push-in connectors used with some petrol engine compression testers cannot be used for diesel engines because of the higher pressures involved. Instead, the diesel engine compression tester screws into an injector or glow plug hole, using one of the adapters supplied with the tester.

Most compression testers are used while cranking the engine on the starter motor. A few can be used with the engine idling; this gives more reliable results, since it is hard to guarantee that cranking speed will not fall in the course of testing all cylinders, whereas the idle speed should remain constant.

Leakdown tester

The leakdown tester measures the rate at which air pressure is lost from each cylinder, and can also be used to pinpoint the source of pressure loss (valves, head gasket or bores). Its use depends on the availability of a supply of compressed air, typically at 5 to 10 bar. The same tester (with different adapters) can be used on both petrol and diesel engines *(see illustration)*.

In use, the tester is connected to an air line and to an adapter screwed into the injector or glow plug hole, with the piston concerned at TDC on the compression stroke. The procedure is described in Chapter 5, Section 4.

Vacuum pump and/or gauge

A vacuum gauge with suitable adapters is useful for locating blockages or inlet air leaks in the supply side of the fuel system. A simple gauge is used with the engine running to create a vacuum in the supply lines. A hand-held vacuum pump with its own gauge can be used without running the engine, and is also useful for bleeding the fuel system when a hand-priming pump is not fitted *(see illustration)*. Test procedures are given in Chapter 5, Section 3.

Crow-foot adapters

With modern common rail injection systems, the pressure of the fuel being injected can be almost 2000 bar (29 400 psi), so the condition of the rigid metal pipes between the high-pressure fuel pump, the accumulator rail and the injectors is critical. If disturbed, the pipes must be replaced, and after fitting new ones the unions must be tightened to the correct

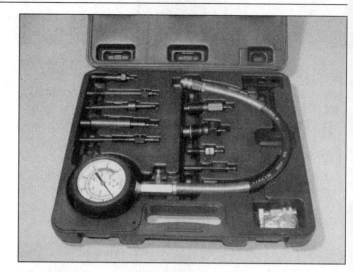

Diesel engine compression tester kit

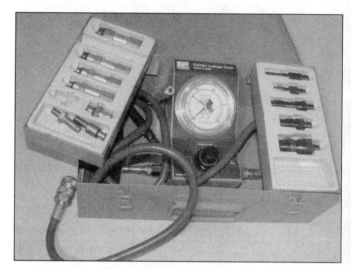

Typical leakdown tester kit

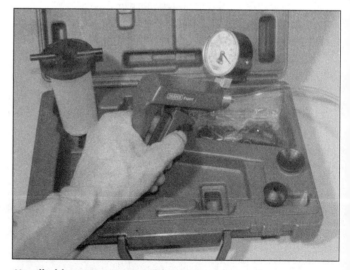

Handheld vacuum pump with gauge

Crow-foot adapters ...

... will be necessary for tightening the fuel pipe unions

torque setting. As a conventional socket cannot be used, a crow-foot adapter should be used on the end of a torque wrench *(see illustrations)*.

3 Injection pump timing tools

If work is undertaken which disturbs the position of the fuel injection pump, certain tools will be required to check the injection pump timing on reassembly. This also applies if the pump drive is disturbed – including renewal of the timing belt on some models. Checking of the timing is also a necessary part of fault diagnosis when investigating complaints such as power loss, knock and smoke. **Note:** *On common rail engines the injection timing is controlled solely be the ECU - the pump simply provides the pressurised fuel. Therefore when refitting the pump, no timing procedure is necessary.*

Static timing tools

Static timing is still the most widely-used method of setting diesel injection pump, but it is time-consuming and sometimes messy. Precision measuring instruments are often needed for dealing with distributor pumps, and good results depend on the skill and patience of the operator.

The owner-mechanic who will only be dealing with one engine should refer to the manufacturer's information or to the relevant Haynes Service and Repair Manual to find out what tools will be required. The diesel engine specialist will typically need the following:

Two dial test indicators (DTI) with magnetic stands.
DTI adapters and probes for Bosch and CAV distributor pumps.
Spill tube for in-line pumps.
Timing gear pins and pegs.
Crankshaft or flywheel locking pins.

Dial test indicator and magnetic stand

This is a useful workshop tool for many operations besides timing: it is, for example, the most accurate means of

checking the protrusion or recess of swirl chambers, pistons and liners when renewing cylinder head gaskets *(see illustrations)*. If major overhauls are undertaken, it can also be used for measuring values such as crankshaft endfloat.

A dial test indicator can also be used to check the swirl chamber protrusion ...

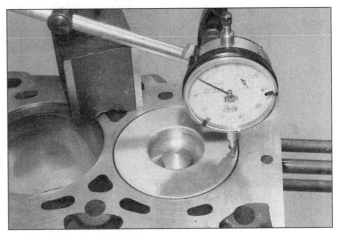

... and piston protrusion

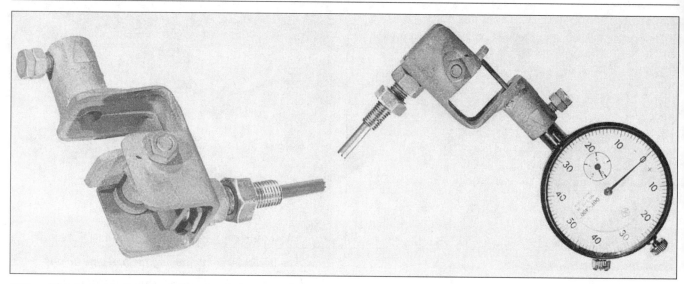

DTI and fabricated bellcrank adapter for timing a Bosch VE pump

Two DTIs are needed for setting the timing on some engines (for instance, the early Peugeot/Citroen XUD series): one to measure the pump plunger or rotor movement, and one to measure engine piston protrusion.

DTI adapters

Adapters and probes for fitting the DTI to the distributor pump are of various patterns, due partly to the need to be able to use them in conditions of poor access on the vehicle *(see illustration)*. This means that the same adapter cannot necessarily be used on the same type of pump and engine if the under-bonnet layout is different. On the bench, it is often possible to use similar equipment.

A spring-loaded probe is required on some CAV/RotoDiesel pumps to find the timing groove in the pump rotor *(see illustration)*.

Spill tube

This is a relatively cheap and simple piece of equipment, used for timing many in-line pumps. The tube is fitted in place of one of the pump delivery valves. The traditional form of tube has a 'swan neck' shape; more modern versions have a transparent vertical tube with a calibrated line. A spill tube can easily be made in the workshop using an old injector pipe *(see illustration)*.

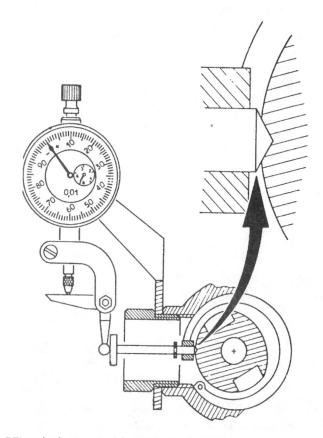

DTI and adapter used for timing Lucas/CAV

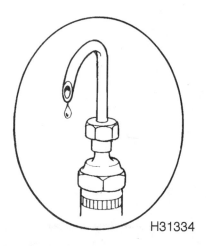

H31334

Simple spill tube

Timing gear pins or pegs

Pins or pegs are used on some engines to lock the pump and/or the camshaft in a particular position. They are generally specific to a particular engine or manufacturer. It is sometimes possible to use suitably-sized dowel rods, drill shanks or bolts instead (see illustration).

Crankshaft or flywheel locking pins

These are used for locking the crankshaft at TDC (or at the injection point on some models). The crankshaft locking pin is inserted through a hole in the side of the crankcase after removal of a plug, and enters a slot in a crankshaft counterweight or web. The flywheel pin passes through a hole in the flywheel end of the crankcase, and enters a hole in the flywheel. Again, suitably-sized rods or bolts can sometimes be used instead (see illustration).

Dynamic timing tools

Dynamic timing on diesel engines is not as widespread as static timing, due partly to the relatively expensive equipment required. Additionally, not all vehicle manufacturers provide dynamic timing values.

Most dynamic timing equipment depends on converting mechanical or hydraulic impulses in the injection system into electrical signals. An alternative approach is to use an optical-to-electrical conversion, with a sensor which screws into a glow plug hole and 'sees' the light of combustion.

Not all diesel engines have ready-made timing marks. If the engine has a TDC sensor, and the timing equipment can read the sensor output, this is not a problem. Some engines have neither timing marks nor TDC sensors; in such cases, there is no choice but to establish TDC accurately, and make suitable marks on the flywheel or crankshaft pulley.

For these reasons, dynamic timing methods and the tools required are not described in this book.

Timing pin (arrowed) or pegs are used to lock the camshaft/injection pump in a particular position

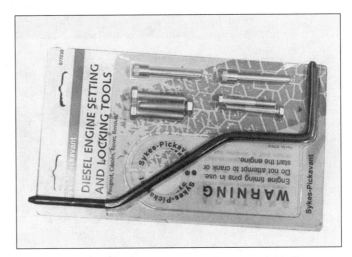

Some crankshaft/flywheel locking tools are available for specific engines (PSA XUD shown) ...

4   Injector testing equipment

> ⊘ *Warning: never expose the hands, face or any other part of the body to injector spray. The high working pressure can penetrate the skin, with potentially fatal results. When possible, use injector test fluid rather than fuel for testing. Take precautions to avoid inhaling the vaporised fuel or injector test fluid. Remember that even diesel fuel is inflammable when vaporised.*

Some kind of injector tester will be needed if it is wished to identify defective injectors, or to test them after cleaning or prolonged storage. Various makes and models are available, but the essential components of all of them are a high-pressure hand-operated pump and a pressure gauge.

... whilst on others, a suitable rod (arrowed) or bolt can be used (Freelander TD4 shown)

Chapter 6

Typical injection tester in use on the bench ...

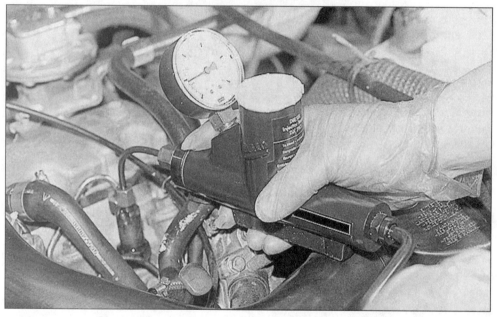

... and on the engine

For safety reasons, injector test or calibration fluid should be used for bench-testing rather than diesel fuel or paraffin *(see illustration)*. Use the fluid specified by the manufacturer of the test equipment if possible.

Some of the simpler testers have the advantage that they can be used to test opening pressure and back-leakage without removing the injectors from the engine *(see illustration)*. A small reservoir may make such testers of limited use for bench-testing, but good results can be obtained with practice.

Another method of testing injectors on the engine is to connect a pressure gauge into the line between the injection pump and the injector. This test can also detect faults caused by the injection pump high-pressure piston or delivery valve.

The workshop which tests or calibrates injectors regularly will need a bench-mounted tester. These testers have a lever-operated pump, and a larger fluid reservoir than the hand-held tester. The best models also incorporate a transparent chamber for safe viewing of the injector spray pattern, and perhaps a test fluid recirculation system *(see illustrations)*.

Some means of extracting the vapour produced when testing, such as a hood connected to the workshop's fume extraction system, is desirable. Although injector test fluid is relatively non-toxic, its vapour is not particularly pleasant to inhale.

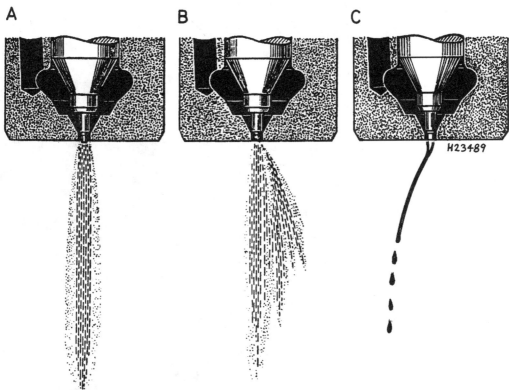

Pintle injector spray patterns

A Good – well-defined spray   B Bad – poorly-defined, ragged spray   C Bad – 'hosing'

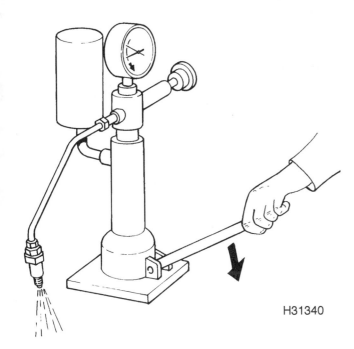

Checking injector spray pattern with an injector tester

5 Injection pump testing and calibration equipment

The equipment needed for testing and calibration of injection pumps is beyond the scope of this book. Any such work should be entrusted to the pump manufacturer's agent – though the opportunity is taken to say yet again that the injection pump is often blamed for faults, when in fact the trouble lies elsewhere.

6 Smoke testing equipment

In most European countries, smoke emission testing is mandatory for heavy goods vehicles, and for passenger vehicles as part of the annual roadworthiness test.

Smoke testing equipment falls into two categories : indirect and direct reading. With the indirect systems, a sample of exhaust gas is passed over a filter paper, and the change in opacity of the paper is measured using a separate machine. With the direct systems, an optically-sensitive probe measures the opacity of the exhaust gas, and an immediate read-out is available.

Chapter 6

7 Electronic fault code readers and scanners

Like their petrol engine counterparts, many of the modern electronic diesel engine control systems have a self-diagnostic function, which continually monitors the operation of the system.

The self-diagnostic system is able to detect system faults such as a faulty sensor or actuator, and can allocate a fault code to identify the source of the problem. The system stores any fault codes in the electronic control unit (ECU) memory, and if a fault is present, a warning light will normally be illuminated on the instrument panel to inform the driver.

Fault codes can be read using a suitable electronic fault code reader/scanner. Most vehicle manufacturer's produce their own dedicated diagnostic equipment, but aftermarket fault code readers/scanners are also available from various manufacturers *(see illustrations)*. The latest diesel engine management ECU self-diagnosis systems are also able to provide 'live data', and system 'snap-shots'. Here the scanner displays the actual output (live data) from each sensor, and the actuator signal from the ECU, as they are generated. In snap-shot mode, the ECU records the values of various sensors at the moment a fault code is generated. These two facilities can provide invaluable assistance in rectifying running faults.

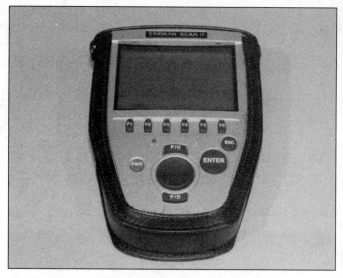

Fault code readers/scanners come in all sorts of shapes ...

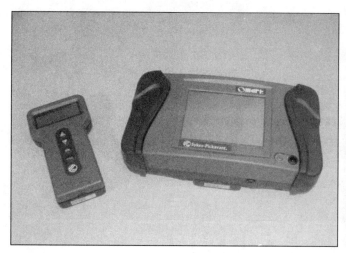

... and sizes

Index

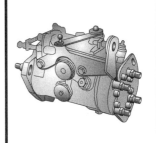

Index

Index

T

V

W

Haynes Manuals – The Complete List

Title	Book No.
ALFA ROMEO Alfasud/Sprint (74 - 88) up to F *	0292
Alfa Romeo Alfetta (73 - 87) up to E *	0531
AUDI 80, 90 & Coupe Petrol (79 - Nov 88) up to F	0605
Audi 80, 90 & Coupe Petrol (Oct 86 - 90) D to H	1491
Audi 100 & 200 Petrol (Oct 82 - 90) up to H	0907
Audi 100 & A6 Petrol & Diesel (May 91 - May 97) H to P	3504
Audi A4 Petrol & Diesel (95 - Feb 00) M to V	3575
AUSTIN A35 & A40 (56 - 67) up to F *	0118
Austin/MG/Rover Maestro 1.3 & 1.6 Petrol (83 - 95) up to M	0922
Austin/MG Metro (80 - May 90) up to G	0718
Austin/Rover Montego 1.3 & 1.6 Petrol (84 - 94) A to L	1066
Austin/MG/Rover Montego 2.0 Petrol (84 - 95) A to M	1067
Mini (59 - 69) up to H	0527
Mini (69 - 01) up to X	0646
Austin/Rover 2.0 litre Diesel Engine (86 - 93) C to L	1857
AUSTIN HEALEY 100/6 & 3000 (56 - 68) up to G *	0049
BEDFORD CF Petrol (69 - 87) up to E	0163
Bedford/Vauxhall Rascal & Suzuki Supercarry (86 - Oct 94) C to M	3015
BMW 316, 320 & 320i (4-cyl) (75 - Feb 83) up to Y *	0276
BMW 320, 320i, 323i & 325i (6-cyl) (Oct 77 - Sept 87) up to E	0815
BMW 3- & 5-Series Petrol (81 - 91) up to J	1948
BMW 3-Series Petrol (Apr 91 - 96) H to N	3210
BMW 3-Series Petrol (Sept 98 - 03) S-reg. on	4067
BMW 520i & 525e (Oct 81 - June 88) up to E	1560
BMW 525, 528 & 528i (73 - Sept 81) up to X *	0632
BMW 1500, 1502, 1600, 1602, 2000 & 2002 (59 - 77) up to S *	0240
CHRYSLER PT Cruiser Petrol (00 - 03) W-reg. on	4058
CITROËN 2CV, Ami & Dyane (67 - 90) up to H	0196
Citroën AX Petrol & Diesel (87 - 97) D to P	3014
Citroën BX Petrol (83 - 94) A to L	0908
Citroën C15 Van Petrol & Diesel (89 - Oct 98) F to S	3509
Citroën CX Petrol (75 - 88) up to F	0528
Citroën Saxo Petrol & Diesel (96 - 01) N to X	3506
Citroën Visa Petrol (79 - 88) up to F	0620
Citroën Xantia Petrol & Diesel (93 - 98) K to S	3082
Citroën XM Petrol & Diesel (89 - 00) G to X	3451
Citroën Xsara Petrol & Diesel (97 - Sept 00) R to W	3751
Citroën Xsara Picasso Petrol & Diesel (00 - 02) W-reg. onwards	3944
Citroën ZX Diesel (91 - 98) J to S	1922
Citroën ZX Petrol (91 - 98) H to S	1881
Citroën 1.7 & 1.9 litre Diesel Engine (84 - 96) A to N	1379
FIAT 126 (73 - 87) up to E *	0305
Fiat 500 (57 - 73) up to M *	0090
Fiat Bravo & Brava Petrol (95 - 00) N to W	3572
Fiat Cinquecento (93 - 98) K to R	3501
Fiat Panda (81 - 95) up to M	0793
Fiat Punto Petrol & Diesel (94 - Oct 99) L to V	3251
Fiat Punto Petrol (Oct 99 - July 03) V-reg on	4066
Fiat Regata Petrol (84 - 88) A to F	1167
Fiat Tipo Petrol (88 - 91) E to J	1625
Fiat Uno Petrol (83 - 95) up to M	0923
Fiat X1/9 (74 - 89) up to G *	0273
FORD Anglia (59 - 68) up to G *	0001

Title	Book No.
Ford Capri II (& III) 1.6 & 2.0 (74 - 87) up to E	0283
Ford Capri II (& III) 2.8 & 3.0 V6 (74 - 87) up to E	1309
Ford Cortina Mk III 1300 & 1600 (70 - 76) up to P*	0070
Ford Escort Mk I 1100 & 1300 (68 - 74) up to N*	0171
Ford Escort Mk I Mexico, RS 1600 & RS 2000 (70 - 74) up to N *	0139
Ford Escort Mk II Mexico, RS 1800 & RS 2000 (75 - 80) up to W *	0735
Ford Escort (75 - Aug 80) up to V *	0280
Ford Escort Petrol (Sept 80 - Sept 90) up to H	0686
Ford Escort & Orion Petrol (Sept 90 - 00) H to X	1737
Ford Escort & Orion Diesel (Sept 90 - 00) H to X	4081
Ford Fiesta (76 - Aug 83) up to Y	0334
Ford Fiesta Petrol (Aug 83 - Feb 89) A to F	1030
Ford Fiesta Petrol (Feb 89 - Oct 95) F to N	1595
Ford Fiesta Petrol & Diesel (Oct 95 - 01) N-reg. on	3397
Ford Fiesta (02 - 04) 02-reg. onwards	4170
Ford Focus Petrol & Diesel (98 - 01) S to Y	3759
Ford Focus Petrol & Diesel (01 - 04) Y-reg. on	4167
Ford Galaxy Petrol & Diesel (95 - Aug 00) M to W	3984
Ford Granada Petrol (Sept 77 - Feb 85) up to B	0481
Ford Granada & Scorpio Petrol (Mar 85 - 94) B to M	1245
Ford Ka (96 - 02) P-reg. onwards	3570
Ford Mondeo Petrol (93 - Sept 00) K to X	1923
Ford Mondeo Petrol & Diesel (Oct 00 - Jul 03) X to 03	3990
Ford Mondeo Diesel (93 - 96) L to N	3465
Ford Orion Petrol (83 - Sept 90) up to H	1009
Ford Sierra 4-cyl Petrol (82 - 93) up to K	0903
Ford Sierra V6 Petrol (82 - 91) up to J	0904
Ford Transit Petrol (Mk 2) (78 - Jan 86) up to C	0719
Ford Transit Petrol (Mk 3) (Feb 86 - 89) C to G	1468
Ford Transit Diesel (Feb 86 - 99) C to T	3019
Ford 1.6 & 1.8 litre Diesel Engine (84 - 96) A to N	1172
Ford 2.1, 2.3 & 2.5 litre Diesel Engine (77 - 90) up to H	1606
FREIGHT ROVER Sherpa Petrol (74 - 87) up to E	0463
HILLMAN Avenger (70 - 82) up to Y	0037
Hillman Imp (63 - 76) up to R *	0022
HONDA Accord (76 - Feb 84) up to A	0351
Honda Civic (Feb 84 - Oct 87) A to E	1226
Honda Civic (Nov 91 - 96) J to N	3199
Honda Civic Petrol (Mar 95 - 00) M to X	4050
HYUNDAI Pony (85 - 94) C to M	3398
JAGUAR E Type (61 - 72) up to L	0140
Jaguar MkI & II, 240 & 340 (55 - 69) up to H *	0098
Jaguar XJ6, XJ & Sovereign; Daimler Sovereign (68 - Oct 86) up to D	0242
Jaguar XJ6 & Sovereign (Oct 86 - Sept 94) D to M	3261
Jaguar XJ12, XJS & Sovereign; Daimler Double Six (72 - 88) up to F	0478
JEEP Cherokee Petrol (93 - 96) K to N	1943
LADA 1200, 1300, 1500 & 1600 (74 - 91) up to J	0413
Lada Samara (87 - 91) D to J	1610
LAND ROVER 90, 110 & Defender Diesel (83 - 95) up to N	3017
Land Rover Discovery Petrol & Diesel (89 - 98) G to S	3016
Land Rover Freelander Petrol & Diesel (97 - 02) R-reg. onwards	3929
Land Rover Series IIA & III Diesel (58 - 85) up to C	0529
Land Rover Series II, IIA & III 4-cyl Petrol (58 - 85) up to C	0314
MAZDA 323 (Mar 81 - Oct 89) up to G	1608

Title	Book No.
Mazda 323 (Oct 89 - 98) G to R	3455
Mazda 626 (May 83 - Sept 87) up to E	0929
Mazda B-1600, B-1800 & B-2000 Pick-up Petrol (72 - 88) up to F	0267
Mazda RX-7 (79 - 85) up to C *	0460
MERCEDES-BENZ 190, 190E & 190D Petrol & Diesel (83 - 93) A to L	3450
Mercedes-Benz 200 D, 240 D, 240 TD, 300 D & 300 TD 123 Series Diesel (Oct 76 - 85) up to C	1114
Mercedes-Benz 250 & 280 (68 - 72) up to L	0346
Mercedes-Benz 250 & 280 123 Series Petrol (Oct 76 - 84) up to B *	0677
Mercedes-Benz 124 Series Petrol & Diesel (85 - Aug 93) C to K	3253
Mercedes-Benz C-Class Petrol & Diesel (93 - Aug 00) L to W	3511
MG A (55 - 62) *	0475
MGB (62 - 80) up to W	0111
MG Midget & Austin-Healey Sprite (58 - 80) up to W	0265
MITSUBISHI Shogun & L200 Pick-Ups Petrol (83 - 94) up to M	1944
MORRIS Ital 1.3 (80 - 84) up to B	0705
Morris Minor 1000 (56 - 71) up to K	0024
NISSAN Almera Petrol (95 - Feb 00) N to V	4053
Nissan Bluebird (May 84 - Mar 86) A to C	1223
Nissan Bluebird Petrol (Mar 86 - 90) C to H	1473
Nissan Cherry (Sept 82 - 86) up to D	1031
Nissan Micra (83 - Jan 93) up to K	0931
Nissan Micra (93 - 99) K to T	3254
Nissan Primera Petrol (90 - Aug 99) H to T	1851
Nissan Stanza (82 - 86) up to D	0824
Nissan Sunny Petrol (May 82 - Oct 86) up to D	0895
Nissan Sunny Petrol (Oct 86 - Mar 91) D to H	1378
Nissan Sunny Petrol (Apr 91 - 95) H to N	3219
OPEL Ascona & Manta (B Series) (Sept 75 - 88) up to F	0316
Opel Ascona Petrol (81 - 88) (Not available in UK see Vauxhall Cavalier 0812)	3215
Opel Astra Petrol (Oct 91 - Feb 98) (Not available in UK see Vauxhall Astra 1832)	3156
Opel Astra & Zafira Diesel (Feb 98 - Sept 00) (See Vauxhall/Opel Astra & Zafira Diesel Book No. 3797)	
Opel Astra & Zafira Petrol (Feb 98 - Sept 00) (See Vauxhall/Opel Astra & Zafira Petrol Book No. 3758)	
Opel Calibra (90 - 98) (See Vauxhall/Opel Calibra Book No. 3502)	
Opel Corsa Petrol (83 - Mar 93) (Not available in UK see Vauxhall Nova 0909)	3160
Opel Corsa Petrol (Mar 93 - 97) (Not available in UK see Vauxhall Corsa 1985)	3159
Opel Corsa Diesel (Mar 93 - Oct 00) (See Vauxhall/Opel Corsa Diesel Book No. 4087)	
Opel Corsa Petrol (Apr 97 - Oct 00) (See Vauxhall/Opel Corsa Petrol Book No. 3921)	
Opel Corsa Petrol & Diesel (Oct 00 - Sept 03) (See Vauxhall/Opel Corsa Petrol & Diesel Book No. 4079)	
Opel Frontera Petrol & Diesel (91 - 98) (See Vauxhall/Opel Frontera Book No. 3454)	
Opel Kadett Petrol (Nov 79 - Oct 84) up to B	0634
Opel Kadett Petrol (Oct 84 - Oct 91) (Not available in UK see Vauxhall Astra & Belmont 1136)	3196
Opel Omega & Senator Petrol (Nov 86 - 94) (NA in UK see Vauxhall Carlton & Senator 1469)	3157
Opel Omega (94 - 99) (See Vauxhall/Opel Omega Book No. 3510)	
Opel Rekord Petrol (Feb 78 - Oct 86) up to D	0543

* Classic reprint

Title	Book No.
Opel Vectra Petrol (Oct 88 - Oct 95)	3158
(Not available in UK see Vauxhall Cavalier 1570)	
Opel Vectra (95 - Feb 99)	
(See Vauxhall/Opel Vectra Book No. 3396)	
Opel Vectra (Mar 99 - May 02)	
(See Vauxhall/Opel Vectra Book No. 3930)	
Opel Diesel Engine *(See Vauxhall/Opel 1.5, 1.6 & 1.7 litre Diesel Engine Book No. 1222)*	
PEUGEOT 106 Petrol & Diesel (91 - 02) J-reg. on	1882
Peugeot 205 Petrol (83 - 97) A to P	0932
Peugeot 206 Petrol & Diesel (98 - 01) S to X	3757
Peugeot 306 Petrol & Diesel (93 - 99) K to T	3073
Peugeot 307 Petrol & Diesel (01 - 04) Y-reg. on	4147
Peugeot 309 Petrol (86 - 93) C to K	1266
Peugeot 405 Petrol (88 - 97) E to P	1559
Peugeot 405 Diesel (88 - 97) E to P	3198
Peugeot 406 Petrol & Diesel (96 - Mar 99) N to T	3394
Peugeot 406 Petrol & Diesel (Mar 99 - 02) T-reg. onwards	3982
Peugeot 505 Petrol (79 - 89) up to G	0762
Peugeot 1.7/1.8 & 1.9 litre Diesel Engine (82 - 96) up to N	0950
Peugeot 2.0, 2.1, 2.3 & 2.5 litre Diesel Engines (74 - 90) up to H	1607
PORSCHE 911 (65 - 85) up to C	0264
Porsche 924 & 924 Turbo (76 - 85) up to C	0397
PROTON (89 - 97) F to P	3255
RANGE ROVER V8 Petrol (70 - Oct 92) up to K	0606
RELIANT Robin & Kitten (73 - 83) up to A *	0436
RENAULT 4 (61 - 86) up to D *	0072
Renault 5 Petrol (Feb 85 - 96) B to N	1219
Renault 9 & 11 Petrol (82 - 89) up to F	0822
Renault 18 Petrol (79 - 86) up to D	0598
Renault 19 Petrol (89 - 96) F to N	1646
Renault 19 Diesel (89 - 96) F to N	1946
Renault 21 Petrol (86 - 94) C to M	1397
Renault 25 Petrol & Diesel (84 - 92) B to K	1228
Renault Clio Petrol (91 - May 98) H to R	1853
Renault Clio Diesel (91 - June 96) H to N	3031
Renault Clio Petrol & Diesel (May 98 - May 01) R to Y	3906
Renault Clio Petrol & Diesel (June 01 - 04) Y-reg. onwards	4168
Renault Espace Petrol & Diesel (85 - 96) C to N	3197
Renault Laguna Petrol & Diesel (94 - 00) L to W	3252
Renault Mégane & Scénic Petrol & Diesel (96 - 98) N to R	3395
Renault Mégane & Scénic Petrol & Diesel (Apr 99 - 02) T-reg. onwards	3916
ROVER 213 & 216 (84 - 89) A to G	1116
Rover 214 & 414 Petrol (89 - 96) G to N	1689
Rover 216 & 416 Petrol (89 - 96) G to N	1830
Rover 211, 214, 216, 218 & 220 Petrol & Diesel (Dec 95 - 98) N to R	3399
Rover 25 & MG ZR Petrol & Diesel (Oct 99 - 03) V-reg. onwards	4145
Rover 414, 416 & 420 Petrol & Diesel (May 95 - 98) M to R	3453
Rover 618, 620 & 623 Petrol (93 - 97) K to P	3257
Rover 820, 825 & 827 Petrol (86 - 95) D to N	1380
Rover 3500 (76 - 87) up to E *	0365
Rover Metro, 111 & 114 Petrol (May 90 - 98) G to S	1711
SAAB 95 & 96 (66 - 76) up to R *	0198
Saab 90, 99 & 900 (79 - Oct 93) up to L	0765
Saab 900 (Oct 93 - 98) L to R	3512
Saab 9000 (4-cyl) (85 - 98) C to S	1686

Title	Book No.
Saab 9-5 Petrol (Sept 97 - 03) R-reg. onwards	4156
SEAT Ibiza & Cordoba Petrol & Diesel (Oct 93 - Oct 99) L to V	3571
Seat Ibiza & Malaga Petrol (85 - 92) B to K	1609
SKODA Estelle (77 - 89) up to G	0604
Skoda Favorit (89 - 96) F to N	1801
Skoda Felicia Petrol & Diesel (95 - 01) M to X	3505
SUBARU 1600 & 1800 (Nov 79 - 90) up to H	0995
SUNBEAM Alpine, Rapier & H120 (67 - 74) up to N *	0051
SUZUKI SJ Series, Samurai & Vitara (4-cyl) Petrol (82 - 97) up to P	1942
TALBOT Alpine, Solara, Minx & Rapier (75 - 86) up to D	0337
Talbot Horizon Petrol (78 - 86) up to D	0473
Talbot Samba (82 - 86) up to D	0823
TOYOTA Carina E Petrol (May 92 - 97) J to P	3256
Toyota Corolla (80 - 85) up to C	0683
Toyota Corolla (Sept 83 - Sept 87) A to E	1024
Toyota Corolla (Sept 87 - Aug 92) E to K	1683
Toyota Corolla Petrol (Aug 92 - 97) K to P	3259
Toyota Hi-Ace & Hi-Lux Petrol (69 - Oct 83) up to A	0304
TRIUMPH GT6 & Vitesse (62 - 74) up to N *	0112
Triumph Herald (59 - 71) up to K *	0010
Triumph Spitfire (62 - 81) up to X	0113
Triumph Stag (70 - 78) up to T *	0441
Triumph TR2, TR3, TR3A, TR4 & TR4A (52 - 67) up to F *	0028
Triumph TR5 & 6 (67 - 75) up to P *	0031
Triumph TR7 (75 - 82) up to Y *	0322
VAUXHALL Astra Petrol (80 - Oct 84) up to B	0635
Vauxhall Astra & Belmont Petrol (Oct 84 - Oct 91) B to J	1136
Vauxhall Astra Petrol (Oct 91 - Feb 98) J to R	1832
Vauxhall/Opel Astra & Zafira Petrol (Feb 98 - Sept 00) R to W	3758
Vauxhall/Opel Astra & Zafira Diesel (Feb 98 - Sept 00) R to W	3797
Vauxhall/Opel Calibra (90 - 98) G to S	3502
Vauxhall Carlton Petrol (Oct 78 - Oct 86) up to D	0480
Vauxhall Carlton & Senator Petrol (Nov 86 - 94) D to L	1469
Vauxhall Cavalier Petrol (81 - Oct 88) up to F	0812
Vauxhall Cavalier Petrol (Oct 88 - 95) F to N	1570
Vauxhall Chevette (75 - 84) up to B	0285
Vauxhall/Opel Corsa Diesel (Mar 93 - Oct 00) K to X	4087
Vauxhall Corsa Petrol (Mar 93 - 97) K to R	1985
Vauxhall/Opel Corsa Petrol (Apr 97 - Oct 00) P to X	3921
Vauxhall/Opel Corsa Petrol & Diesel (Oct 00 - Sept 03) X-reg onwards	4079
Vauxhall/Opel Frontera Petrol & Diesel (91 - Sept 98) J to S	3454
Vauxhall Nova Petrol (83 - 93) up to K	0909
Vauxhall/Opel Omega Petrol (94 - 99) L to T	3510
Vauxhall/Opel Vectra Petrol & Diesel (95 - Feb 99) N to S	3396
Vauxhall/Opel Vectra Petrol & Diesel (Mar 99 - May 02) T-reg. onwards	3930
Vauxhall/Opel 1.5, 1.6 & 1.7 litre Diesel Engine (82 - 96) up to N	1222
VOLKSWAGEN 411 & 412 (68 - 75) up to P *	0091
Volkswagen Beetle 1200 (54 - 77) up to S	0036
Volkswagen Beetle 1300 & 1500 (65 - 75) up to P	0039
Volkswagen Beetle 1302 & 1302S (70 - 72) up to L *	0110

Title	Book No.
Volkswagen Beetle 1303, 1303S & GT (72 - 75) up to P	0159
Volkswagen Beetle Petrol & Diesel (Apr 99 - 01) T-reg. onwards	3798
Volkswagen Golf & Bora Petrol & Diesel (April 98 - 00) R to X	3727
Volkswagen Golf & Jetta Mk 1 Petrol 1.1 & 1.3 (74 - 84) up to A	0716
Volkswagen Golf, Jetta & Scirocco Mk 1 Petrol 1.5, 1.6 & 1.8 (74 - 84) up to A	0726
Volkswagen Golf & Jetta Mk 1 Diesel (78 - 84) up to A	0451
Volkswagen Golf & Jetta Mk 2 Petrol (Mar 84 - Feb 92) A to J	1081
Volkswagen Golf & Vento Petrol & Diesel (Feb 92 - Mar 98) J to R	3097
Volkswagen Golf (01 - 04) X-reg. onwards	4169
Volkswagen LT Petrol Vans & Light Trucks (76 - 87) up to E	0637
Volkswagen Passat & Santana Petrol (Sept 81 - May 88) up to E	0814
Volkswagen Passat 4-cyl Petrol & Diesel (May 88 - 96) E to P	3498
Volkswagen Passat 4-cyl Petrol & Diesel (Dec 96 - Nov 00) P to X	3917
Volkswagen Polo & Derby (76 - Jan 82) up to X	0335
Volkswagen Polo (82 - Oct 90) up to H	0813
Volkswagen Polo Petrol (Nov 90 - Aug 94) H to L	3245
Volkswagen Polo Hatchback Petrol & Diesel (94 - 99) M to S	3500
Volkswagen Scirocco (82 - 90) up to H	1224
Volkswagen Transporter 1600 (68 - 79) up to V	0082
Volkswagen Transporter 1700, 1800 & 2000 (72 - 79) up to V *	0226
Volkswagen Transporter (air-cooled) Petrol (79 - 82) up to Y	0638
Volkswagen Transporter (water-cooled) Petrol (82 - 90) up to H	3452
Volkswagen Type 3 (63 - 73) up to M *	0084
VOLVO 120 & 130 Series (& P1800) (61 - 73) up to M *	0203
Volvo 142, 144 & 145 (66 - 74) up to N *	0129
Volvo 240 Series Petrol (74 - 93) up to K	0270
Volvo 262, 264 & 260/265 (75 - 85) up to C *	0400
Volvo 340, 343, 345 & 360 (76 - 91) up to J	0715
Volvo 440, 460 & 480 Petrol (87 - 97) D to P	1691
Volvo 740 & 760 Petrol (82 - 91) up to J	1258
Volvo 850 Petrol (92 - 96) J to P	3260
Volvo 940 Petrol (90 - 96) H to N	3249
Volvo S40 & V40 Petrol (96 - 99) N to V	3569
Volvo S70, V70 & C70 Petrol (96 - 99) P to V	3573

AUTOMOTIVE TECHBOOKS

Title	Book No.
Automotive Air Conditioning Systems	3740
Automotive Carburettor Manual	3288
Automotive Diagnostic Fault Codes Manual	3472
Automotive Diesel Engine Service Guide	3286
Automotive Electrical and Electronic Systems Manual	3049
Automotive Engine Management and Fuel Injection Systems Manual	3344
Automotive Gearbox Overhaul Manual	3473
Automotive Service Summaries Manual	3475
Automotive Timing Belts Manual - Austin/Rover	3549
Automotive Timing Belts Manual - Ford	3474
Automotive Timing Belts Manual - Peugeot/Citroën	3568
Automotive Timing Belts Manual - Vauxhall/Opel	3577
Automotive Welding Manual	3053

CL16.03/04

Preserving Our Motoring Heritage

> <
> *The Model J Duesenberg Derham Tourster. Only eight of these magnificent cars were ever built – this is the only example to be found outside the United States of America*

Almost every car you've ever loved, loathed or desired is gathered under one roof at the Haynes Motor Museum. Over 300 immaculately presented cars and motorbikes represent every aspect of our motoring heritage, from elegant reminders of bygone days, such as the superb Model J Duesenberg to curiosities like the bug-eyed BMW Isetta. There are also many old friends and flames. Perhaps you remember the 1959 Ford Popular that you did your courting in? The magnificent 'Red Collection' is a spectacle of classic sports cars including AC, Alfa Romeo, Austin Healey, Ferrari, Lamborghini, Maserati, MG, Riley, Porsche and Triumph.

A Perfect Day Out

Each and every vehicle at the Haynes Motor Museum has played its part in the history and culture of Motoring. Today, they make a wonderful spectacle and a great day out for all the family. Bring the kids, bring Mum and Dad, but above all bring your camera to capture those golden memories for ever. You will also find an impressive array of motoring memorabilia, a comfortable 70 seat video cinema and one of the most extensive transport book shops in Britain. The Pit Stop Cafe serves everything from a cup of tea to wholesome, home-made meals or, if you prefer, you can enjoy the large picnic area nestled in the beautiful rural surroundings of Somerset.

> *John Haynes O.B.E., Founder and Chairman of the museum at the wheel of a Haynes Light 12.* >

> <
> *Graham Hill's Lola Cosworth Formula 1 car next to a 1934 Riley Sports.*

The Museum is situated on the A359 Yeovil to Frome road at Sparkford, just off the A303 in Somerset. It is about 40 miles south of Bristol, and 25 minutes drive from the M5 intersection at Taunton.
Open 9.30am - 5.30pm (10.00am - 4.00pm Winter) 7 days a week, *except Christmas Day, Boxing Day and New Years Day*
Special rates available for schools, coach parties and outings Charitable Trust No. 292048